Global Software Engineering

Virtualization and Coordination

Global Software Engineering

Virtualization and Coordination

Gamel O. Wiredu

CRC Press
Taylor & Francis Group
Boca Raton London New York

CRC Press is an imprint of the
Taylor & Francis Group, an **informa** business

CRC Press
Taylor & Francis Group
6000 Broken Sound Parkway NW, Suite 300
Boca Raton, FL 33487-2742

First issued in paperback 2022

ISBN 13: 978-1-03-247542-4 (pbk)
ISBN 13: 978-0-367-18481-0 (hbk)

DOI: 10.1201/9780429196591

**Visit the Taylor & Francis Web site at
http://www.taylorandfrancis.com**

**and the CRC Press Web site at
http://www.crcpress.com**

To
Kwaku Sefa Wiredu

Contents

Preface

That technology and organization co-evolve is not a reality and theory that are in question. For there is no doubt that the organization of human activities has been a reflection of technological innovation, change, and use, and vice versa. A typical instance is the coordination of global software engineering (GSE) – an information and communication technology (ICT) innovation underpinned by a preponderance of ICT systems in a globally distributed organizational mode. As technology preponderance enables GSE, organization leads to enhanced technological innovation. Hence, it is strange that the written record on this co-evolution in GSE only addresses organization at the expense of technology.

Organization has been addressed overwhelmingly in GSE coordination through its longstanding logics of rationality (i.e., the efficiency ethos, mechanical methods, and mathematical analysis) and indeterminacy (i.e., effectiveness ethos, natural methods, and functional analysis). The endurance of these logics through the years seems to have generated the illusion that existing literature on GSE can provide us with adequate accounts of the coordination role of ICT. But GSE itself is an organizational mode that is technology-begotten, technology-dominated, and technology-driven – likewise its coordination. It is a direct reflection of ICT innovation, change, and use, yet technology has not yet been made to speak to it. Organizational speech alone therefore distorts the story of co-evolution. Consequently, we currently have low technology explanations and/or fragmented explanations of GSE coordination (via ICT, organization, geography, and information perspectives).

This book addresses technology in GSE coordination as an attempt to balance the story. Thus, I present a different analysis of how and why ICT is implicated in this coordination. I have taken a technological view of GSE coordination, considered the existing fragmented explanations and perspectives, asked new questions about them, and found that technology speaks better through the logic of virtuality (i.e., creativity ethos, electrical methods, and technological analysis) than rationality and indeterminacy. The book proposes this logic as the primary and most appropriate approach for a comprehensive study of GSE coordination. Then it presents a technological and integrated explanation of GSE coordination with the core argument that this coordination is achieved through ICT connectivity and capitalization.

Gamel O. Wiredu
Accra, Ghana

Acknowledgments

It was at the University of Limerick, Ireland, that I began studying the subject of coordination in GSE with Liam Bannon, Daniel Sullivan, Gabriela Avram, Anders Sigfridsson, Anne Sheehan, Michael Hales, Michael Cook, Luigina Ciolfi, Brian Fitzgerald, Pär Ågerfalk, and Helena Holmström Olsson. It is Liam Bannon who invited me to join the Social, Organizational, and Cultural Aspects of Global Software Development (socGSD) Group for my post-doctoral fellowship at Limerick in March 2005. We held several workshops on global software development from which I gained many insights into this organizational mode and different approaches to its study, and I am indebted to them. I also thank the members of the global software team with whom I interacted during my data collection in their company (distributed between the Republic of Ireland and the USA).

socGSD was one of the LERO (the Irish Software Engineering Research Institute) cluster projects funded under the Principal Investigator Grant 03/IN3/1408C from the Science Foundation of Ireland. Through this fund, I had financial support for every aspect of my post-doctoral fellowship, including empirical work in the USA as well as the sponsorship to attend and present aspects of this study at international conferences in France and China. Therefore, I am very grateful to the government and people of the Republic of Ireland and the Science Foundation Ireland in particular for such invaluable support.

When I joined the Ghana Institute of Management and Public Administration (GIMPA) after Limerick, I expanded the scope of my study of this subject and changed my approach from organization to technology. In the development of this technological explanation of coordination, I have been encouraged greatly by my colleague Samuel K. Bonsu, Professor of Marketing. Daniel N. Treku, Kofi Arhin, Raphael Amponsah, Solomon Odei-Appiah, Lovestone Mamattah, Emmanuel Djaba, and Nana Kwame Amagyei, who were my research students at GIMPA have greatly encouraged me to persist in further studying this subject and in writing this book. The interest they took and the aptitudes they developed in research during our numerous research seminar series that I facilitated encouraged me to maintain my own strong interest in research in a relatively weak research culture.

I also thank the review and senior acquisitions editors of Taylor & Francis Group, the publisher of this book, for helpful comments on its earlier drafts.

My mother, Comfort Appiah, who inspired me to write my first book, *Mobile Computer Usability* (2014, Springer) with the question "When are you going to write a book?" came up with another question a few years after I gave her a copy: "Are you writing another book?" And I answered yes, because I had been expecting that question and was writing this book to answer her affirmatively. She and my father, Kwaku Sefa (to whom this book is dedicated), laid the foundation for my development into a man of letters by pressing and encouraging me to read and read and read when I was a small boy. She made me read, then she also read to me; and I mostly remember these Bible verses she taught me: "Thy word is a lamp unto my feet, and a light unto my path" (Psalm 119:105); "They that trust in the LORD shall be as mount Zion, which cannot be removed, but abideth for ever" (Psalm 125:1); and "My son, if sinners entice thee, consent thou not" (Proverbs 1:10). I am sure she has God's eternal blessings through Jesus Christ our Lord and Savior.

My wife, Akosua Nyanta, and our daughter, Sarah Twumwaa, have been my closest companions during the writing of this book. They have provided me with great familial and emotional support that gave me a the sound body and mind I needed for reading, thinking, and writing. One is always tempted to take such close companions for granted and treat their companionship and support with presumption. But I have learned to overcome such a temptation by conjecturing how deficient I would be without them and their unique support. Day by day, they assure me of their love and interest in my academic work, even when they remotely understand the technical contents thereof. I thank them for putting up with me while I was laboring to complete this book.

But God is my true provider, preserver, and helper; the fount of every one of my innumerable blessings; my creator and my redeemer. It is He who has provided me with spiritual inspiration to make this knowledge a contribution toward human progress. And so I say, Bless the LORD, O my soul: and all that is within me, bless his holy name. Bless the LORD, O my soul, and forget not all His benefits.

Chapter 1

Introduction

By virtue of innovation and the ubiquity of information and communication technology (ICT), global distribution of software development resources and activities has emerged as a global virtual work configuration – labeled as global software engineering (GSE). ICT is the main enabler of this configuration because of its power and promise to reduce spatial, temporal, and structural barriers. In organizations where this promise has been fulfilled fully or to a significant degree, they enjoy advantages such as cost reduction, access to global pools of expertise, formulation of global virtual teams, and closer-to-market and round-the-clock development. Eric Overby's process virtualization theory[1] suggests that ICT is the predominant resource that enables organizations to virtualize teamwork configurations and processes across global spaces, time, and structures. The theory assumes that once technology can remove time and space barriers to enable process participation and when technology can authenticate the people participating in the process and track their activities, then, given the absence of the dependent variables, organizations will seek to achieve processes virtualization.

Virtualization in the global context (hereafter, virtuality or virtualization) therefore describes how ICT, in combination with and in transformation of space, place, and other organizational resources, enables spanning of discontinuities between geography, time, organization, culture, and work practices. Yet virtuality is at the same time confronted by the reality of geographic dispersion, electronic dependence, structural dynamism, and national diversity that may hinder team effectiveness in this configuration. Thus, in spite of the advantages provided by technology enablement in GSE, there are theoretical and practical problems of coordination in this work configuration.

Coordination is the management of dependencies, uncertainties, and conflicts between people and activities. Dependency is a goal-relevant relationship between two or more people or tasks in which a task cannot begin or be completed until another one has occurred, begun, or completed. Uncertainty refers to incomplete information about an organizational phenomenon that makes it difficult to predict its behavior accurately. Conflict exists when people involved in task performance hold discrepant views or have interpersonal incompatibilities.

Coordination is a fundamental problem for collocated organizations, but the problem is escalated in the globally distributed organizational context because of issues such as cultural and national diversity of developers, erratic information exchanges, mutual knowledge problems, politics, increased uncertainties, geographical distance, and technological limitations. The issues are more pronounced in GSE, and they undermine the management of dependencies, uncertainties, and conflicts to increase the difficulty of coordination. Given that the difficulty of coordination increases with project size and complexity,[2] the escalated problem of GSE coordination is very real in its theoretical and practical dimensions.

Yet the continuous organization of GSE around the world is testament to adequate coordination practice by global software teams in the face of virtualization. Both the escalated problem of coordination and how organizations achieve coordination are quite perplexing for improving coordination theory. The perplexity has generated high research interest among scholars, especially in information systems and organization studies research streams. The outcomes of this interest are four separate theoretical perspectives on (or approaches to) GSE coordination – technology, information, geography, and organization – developed to complement extant coordination theory and to inform and guide research and management.

Unfortunately, researchers' particular or separate foci on these perspectives have resulted in fragmented and, hence, inadequate, explanations of coordination in existing publications. Fragmented explanations betray the low development of GSE coordination theory that has left us with low understanding of the relationships between these perspectives. Yet the practice of GSE around the world suggests clearly that technology, information, geography, and organization interrelate to provide organizations with cost, access, market, teamwork, and development advantages. This disparity between coordination practice and theory constitutes a significant theoretical gap that this book attempts to fill.

To explain how and why the relationships between these perspectives achieve coordination, this book appraises and integrates them with a virtual approach. The appraisal and integration are necessary for the improvement of GSE coordination theory to account for virtuality of software engineering teamwork. The logic of virtuality is given primacy ahead of the traditional predominant logics of rationality and indeterminacy because, first, it is assumed to be the primary logic of GSE. Second, it encapsulates technology, information, geography, and organization (the four perspectives) and their interrelations better than rationality and indeterminacy. Third, it assumes that ICT is essential and generative. These reasons imply that the best approach to manage a GSE organization is to conceptualize it as an ICT-based representation and simulation of its geography, people, and processes. In short, the appraisal and integration task of this book is approached with the premise that virtuality is the primary essential logic of this organizational configuration.

When we approach the study of the organization, technology, information, and geography of coordination with virtuality, these four perspectives become significant for the development of a technology-centered yet integrated explanation of GSE coordination. The view of ICT in the existing and traditional approaches as instrumental is a major weakness of existing coordination theory since it quite overlooks the essential role ICT plays as a core phenomenon that enables virtual teamwork. Such an assumption and its source (that virtuality is not the primary logic of GSE organization, but rationality and indeterminacy are) must be viewed indeed as very limited. This is because rationality and indeterminacy are epiphenomena that are incapable of reaching into the depths of how the combination of organization, technology, information, and geography are implicated in coordination. By assuming that virtuality is primary but only secondary to coordination, we recognize only its contextual side. Yet it is in its epistemological side – that is, when it is treated as the primary logic of organization – that we can develop a sound technological explanation of GSE coordination. Chapter 3 presents an elaborate explanation of how and why virtuality is the primary logic of GSE organization.

This book proposes a new explanation of GSE coordination practice, showing how and why the virtual logic premise and the four perspectives constitute an improved theory of technology coordination. The new explanation points to a more holistic understanding of how and why ICT is the essence of GSE coordination. The four-fold argument in support of the proposition is this: technology is the material basis of coordination; it is essential for exploitation of geography; information management is achieved by combining technology and human agencies; and technology is essential for the spatial and temporal resolution of the paradox of organization.

In Chapter 2, the motivation for writing this book is justified. It presents a review of the limitations with extant GSE coordination theory to point out the neglect of a virtualization approach to the theory. The review is enlarged to cover existing perspectives of technology, information, geography, and organization. Each of the perspectives is discussed to show that it is a particular aspect of virtuality, and yet limited on its own unless they are integrated. The chapter argues that the lack of a coordination theory that is integrated through a virtualization approach is due to the assumption that virtuality is incidental or contextual rather than essential to GSE. On the whole, the chapter prepares the motivational and theoretical grounds for the subsequent chapters of the book.

In order to demonstrate how and why virtuality is an essential logic of GSE organization, Chapter 3 begins with a discussion of the prevalent logics of organization – rationality and indeterminacy. The assumptions of rationality and indeterminacy are discussed to suggest that virtualization is currently deemed as a quality of these logics instead of being deemed as a logic in its own right. To justify the book's estimation of virtuality as the primary organizational logic, it is discussed in terms of the peculiar task coordination challenges it addresses, as compared with the task coordination challenges addressed by rationality and indeterminacy.

Chapter 4 discusses theories of social presence, media richness, computer-supported cooperative work, and media synchronicity, leading to the argument that these theories are all represented in media ecology theory. Media ecology theory is therefore used as the overarching framework to explain the variability and materiality of ICT. The explanation suggests that ICT should not be understood as a static medium but as one that has been enriched continually – evidenced by the development of group support systems. Furthermore, in spite of tight coupling between the ICT medium, the message it conveys, the transmitter and the receiver, the analysis focuses on the materiality of the ICT medium. Applying this reasoning to the coordination function of teleconferences in GSE, the chapter shows that they enable software developers' multitasking and ready access to information. These are distinctive functions which are discussed to show that technology is material in the virtuality perspective on coordination.

Chapter 5 explains coordination from an information management perspective. This explanation is based on an analysis of how information management is implicated in handling the emergent, increased, and varied uncertainties that attend GSE. The analysis evaluates mechanical and organic information processing capabilities of technology and human resources and activities to show how they combine to manage the four different sources of uncertainties in this work configuration. The chapter conceptualizes coordination in terms of interrelations between selection of software developers' task-resolving resources; exploitation of their distance-bridging activities that are non-technological; and support for their technology-based interactions.

Existing GSE research publications make some references to place, space, and ICT, but how these factors shape coordination are quite unclear in the information systems development literature. Chapter 6 addresses this limitation through analysis of how coordination is underpinned by the phenomenological reality of place on the one hand, and the general reality of space and technology on the other. Coordination is characterized by workplace formation and identification with ad hoc verbal communications and by dataspace utilization and appropriation for cross-site collaborations.

The purpose of Chapter 7 is to conceptualize the paradox of coordination through a review and interpretation of existing literature. The discussions are premised on the assumption that GSE configurations are characterized by a paradox of organization – direct effects of global distribution and compensations for the effects. The paradox manifests in positive and negative relations between dependencies, uncertainties, and conflicts at the same time. The discussions show that coordination is achieved by managing direct and inverse relationships between multiple dimensions of dependencies, uncertainties, and conflicts.

Using the distinct logic of virtualization as well as the previous four chapters of the book as a premise, Chapter 8 undertakes a dialectical interpretation of the contradictory logics of rationality and indeterminacy to propose a synthesized theory of GSE coordination. Labeled as coordination by technology, this theory is explained as the combined use of digital technology supported by developers' dexterity to manage dependencies. The mechanisms of coordination by technology are ICT connection and capitalization.

Chapter 9 uses the results of a case to illustrate the proposed theory of coordination by technology. The illustration shows how coordination by technology can be used to interpret process data from a longitudinal case study wherein coordination modes by a global virtual team were investigated. The data are not used to test the proposed theory, but to illustrate their usefulness in explaining how the practice of coordination reflects a dialectical process whereby technology constitutes a synthesis of coordination by plans and by mutual adjustments. Furthermore, the illustrative data points are not presented as a representative proportion of a certain population in order to serve as the bases for statistical generalizability of the proposed theory. Rather, the theory depends on empirical processes of necessary relations between GSE, ICT, and developers generated by real structures that serve as the bases for analytical generalization.

The book closes with reflections on the contributions proffered by the virtuality approach and technological explanations in Chapter 10. The theoretical contributions are discussed in relation to previous explanations of coordination to argue for the distinctiveness of this book. This chapter provides concluding arguments on how and why technology is not just an instrumental but an epistemological issue. In these arguments, the claims for novel contributions are discussed in terms of explanatory depth and breadth that the analyses in previous chapters demonstrate. Based on the theoretical contributions, the future research implications and directions, as well as the practical guidelines for managers of GSE organizations and projects, are discussed.

Chapter 2

Coordination Theory

The practice and theory of coordination in software engineering are significant issues for project managers and researchers.[1] Coordination was an underpinning philosophy of Frederick Taylor's scientific management, and the study of it has remained the preoccupation of many organizational researchers since then.[2] Generally, the coordination literature reveals some particularity or divergence in explanations of the concept. While a majority of researchers explain coordination as the management of dependencies, others explain it in terms of managing uncertainties,[3] and others explain it in terms of interpersonal and inter-unit conflict management.[4]

The earliest conceptualizations of coordination are found in analyses of types of dependencies, differentiation and integration,[5] and coordination modes.[6] For instance, James D. Thompson, drawing upon James G. March and Herbert A. Simon, analyzed three types of dependencies – pooled, serial, and reciprocal. In pooled dependence, units in an organization appear to be independent of each other and may not interact with each other; yet, total breakdown occurs when one unit performs inadequately. Serial or sequential dependence reflects a situation in which one unit directly and visibly depends on the outcome of another unit's actions to function, but this also represents linear causality.[7] The cyclical nature of causation makes reciprocal or "mutual" dependence[8] a more appropriate conceptualization. Mutual dependence is a normal feature of any organization, as espoused in Scott D. N. Cook and John Seely Brown's knowledge-knowing interrelationships,[9] and in Gregory Bateson's ecological theory of mind systems.[10]

In large sections of the coordination literature, dependencies are implicitly and explicitly antecedents or consequences of uncertainties. For example, Richard L. Daft and Robert E. Lengel argue that "[i]nterdependence increases uncertainty because action by one department can unexpectedly force adaptation by other departments in the production chain."[11] Others also suggest the significance of uncertainty in the problem of coordination.[12] According to Frances Milliken, uncertainty is "an individual's perceived inability to predict something accurately."[13] This understanding resonates with Andrew Van De Ven and colleagues' argument that as tasks increase in uncertainty, coordination mechanisms and impersonal programming are increasingly substituted for coordination processes such as horizontal communications and group meetings.[14]

Exchange of information between interdependent parties is neither simple nor presumed because the willingness of individuals or groups to exchange information is dependent on their motives, frames of reference,[15] and degrees of differentiation.[16] Contradictory motives or strong differentials in frames of reference between interdependent work units can engender conflicts or supplant information exchange and hence worsen uncertainty. Karen A. Jehn's analysis of conflict types revealed two main types – task conflicts and relationship conflicts.[17] Task-focused conflicts are those directly related to the organizational goals and to procedures for accomplishing those goals, while relationship-focused conflicts relate to interpersonal and emotional disagreements that are not related to those goals. These conflict distinctions are, however, conceptual since, in reality, one can easily lead to another; and as such, both relationship-focused and task-focused ones have direct causal interrelationships with dependencies and uncertainties. For instance, in James D. Thompson's analysis of conflict management, he gives prominence to these conflict drifts and interrelationships.[18] His concept of "latent-role conflict," which resonates with Jehn's relationship-focused conflict, refers to non-organizational roles such as nepotism, favoritism, and patronage which, he argues, can be activated in organizational contexts.

These ideas suggest that dependencies, uncertainties, and conflicts are distinct constructs of coordination that are yet related. For instance, they are distinct because coordination problems may arise even when conflicts are absent,[19] and incentives provided to reduce conflicts may undermine coordination.[20] Yet they are also related because when negative relationships between them are ignored, the focus on any one of them generally subsumes the other two, leading to particular and linear conceptualization. This subsumption characterizes the existing literature on coordination in software engineering in general,[21] and global software engineering (GSE) in particular.[22] In these bodies of literature, while there is a general agreement that software development coordination efforts are escalated when development is distributed, approaches to analyses reflect particular conceptualizations.

Perspectives

GSE coordination theory in existing literature has been conceptualized in terms of four dominant perspectives: technology, information, geography, and organization. These perspectives have also served as approaches to coordination theory development in the past under the logics of rationality and indeterminacy. Each of these perspectives is discussed below in terms of its significance, the extent of theoretical contributions made, and related under-researched issues that warrant this book's appraisal and integration of them.

The *technology* perspective is concerned with the functions of various hardware and software applications in computers and networks which are used in GSE. The problem of function is about whether any particular technology is instrumental or essential to coordination. Coordination is defined as managing dependencies between people and activities. Dependencies are real in this task because the distributed organizational structure of engineers reflects the architecture of the software – according to Conway's law which assumes a strong correlation between organization structure and software architecture.[23] Software architecture has different parts in its design. These parts are a reflection of the complexity which is an inherent property of the essence of software engineering.[24] Given that spatial and temporal distances undermine GSE coordination,[25] the information and communication technologies (ICTs) that are installed to manage dependencies in this configuration in order to achieve coordination must be essential technologies beyond instrumental.

There are several types of technology artifacts that are deployed in support of the interactions between people, processes, and information to achieve GSE coordination. On the one hand, there are technologies used for the software engineering task directly such as programming languages, development platforms or environments, bug tracking systems, and knowledge repositories. On the other hand, there are technologies such as various information generation, processing, and interaction systems which enable software engineering to work, and which engineers need for production and productivity.

If we follow Frederick P. Brooks' taxonomy of "essence" and "accidents"[26] of software engineering (essence referring to its construct of interlocking concepts that are natural to the task; accidents to the things that attend to it), then we may describe the both groups of technologies as instrumental; none essential. We may do this because we are concerned with task coordination rather than the task (qua task). We are also concerned with the global version of software development rather than its collocated version. The instrumental and/or essential functions of each of these technologies for coordination are of interest in this book. Of particular interest is the essential function where the technology installed is not just supportive but generative and integral in the achievement of coordination.

However, in extant publications on the technology perspective, researchers view ICT as merely instrumental rather than essential or epistemological in extant GSE coordination theory and practice. The instrumental view shows how extant theories of GSE coordination are quite lacking in ICT explanations but are rich with knowledge,[27] culture,[28] task,[29] temporal distance,[30] and architectural explanations.[31] This is in spite of the fact that all the projects studied had ICT installations for coordination purposes. Because of the instrumental view, there is a distinct scarcity of technological explanations in GSE coordination theory. This gap is mainly due to neglect of how functional affordances of ICT play an essential (not just instrumental) role in coordination. Consequently, we only have an instrumental, and hence inadequate, understanding of the role of ICT in coordination.

If we focus, for instance, on a predominant GSE technology such as a teleconference system (a group support system), we realize that one group of scholars has been researching the functionality of electronic meetings from perspectives apart from coordination, while the other group has been researching coordination from perspectives apart from the functionality of teleconferences (see Appendix 1, Table 2). As each group has not related its research to the other, the result is a gap of understanding concerning the coordination function of electronic meetings. As more organizations are increasingly globalizing their software development activities, the trend has brought new characteristics that give new significance to teleconferences. A teleconference is an electronic meeting, and a meeting is a genre of organizational communication characterized by its own structural, linguistic, and substantive conventions.[32] These conventions differentiate the teleconference from other GSE communication modes such as e-mails, one-to-one phone calls, and instant messages. Therefore, to understand the coordination functions of the teleconference, including its underlying technology, there is the need to analyze it separately as an essential resource and in sufficient detail.

Likewise, this book does not lump the possible modes of communication together as though the relationship between the specific conventions of each communication genre on the one hand and the peculiar characteristics of GSE on the other is not critical to the understanding of its coordination function. On the whole, the technology perspective (the essential role of teleconferences in GSE coordination being just one) is under-researched because of the instrumental view of ICT. Technological explanations of coordination that espouse its essential rather than its instrumental role are quite lacking in existing publications. Against this backdrop, there is the need to appraise the coordination models that define this perspective. Thus, one question this book seeks to address is: *how and why does ICT play an essential role in GSE coordination?* The enablement or generative capacity of ICT indicated in the question is particularly important because this book ultimately aims to develop a complementary theory of technology coordination that gives primacy to ICT. The question is addressed in Chapters 4 and 8.

The *information* perspective is informed by the problem of increased uncertainties in GSE which engender erratic dependencies to undermine coordination. The problem of uncertainties is the problem of information processing, and it is reported in the organizational research literature to be directly related to coordination.[33] In global software projects, uncertainties and information processing are identified with different parameters such as task complexity, task environment, and inter- and intra-team dependencies.[34] However, in the following notable instances of previous research, GSE coordination is explained in terms other than uncertainties.

Rajiv Sabherwal explains coordination of outsourced software development in terms of differences in organizational goals and structures between clients and vendors;[35] James Herbsleb and his associates approach coordination from the perspective of distance, speed, and delay in communications to provide their explanations, and propose resolutions in terms of software architecture plans and communications;[36] Becky Grinter and her associates also proffer methods such as functional areas of expertise, product structure, process steps, and customization as the bases of coordination;[37] Haylan Huang and Eileen Trauth explain it in terms of the impact of diverse cultural perceptions of temporal separation;[38] and Julia Kotlarsky and her associates take a knowledge-based perspective on coordination, arguing that it is more suitable than an information processing perspective in the context of a knowledge-intensive activity such as software development.[39] However, their knowledge-based explanation focuses narrowly on analysis of the intellectual contributions of "coordination mechanisms" at the expense of explanations of relationships between them. Besides, those mechanisms are not grounded in empirical data in their research. Rather, they are a priori concepts that are deduced from the literature. Because of this deduction, their explanation overlooks, for instance, the important issue of expertise sourcing even in their analysis of "social mechanisms."

Each of these instances of GSE coordination theory is incomplete because it addresses partial aspects of the multiple sources of uncertainties that cause erratic interdependencies and undermine coordination. As this book will argue, sources of uncertainties such as knowledge, communication, process steps, and architecture plans do not capture the unpredictable character of software requirements that emerge from diverse global situations. These sources are confined to support for developers' interactions, while important aspects of their selection and exploitation are overlooked. The sources reflect the partial aspects, testifying that less than a holistic approach to the problem of uncertainties has been taken in previous research. This limitation exists mainly because the sources are developed with a priori coordination frameworks. Hence, they miss out on emergent phenomena such as changing software requirements which are very important exemplars of the problem of increased uncertainties. The sources fail to account for interactive and sometimes ambiguous events of GSE which, while they fall outside the scope of extant theories, are critical sources of uncertainties that affect coordination.

As a result of this limitation, extant theories do not facilitate a holistic and grounded analysis of the following research issues: first, the diverse range of uncertainties that undermine GSE coordination; and, second, how the diverse range of human and technological resources in GSE can be mobilized and applied to achieve coordination. In short, despite the fact that most of the problems reported in GSE are fundamentally problems of increased and emergent uncertainties, they have not been given explicit attention in previous research. Thus, another important question addressed by this book is: *in the face of increased and emergent uncertainties, how is information managed to achieve GSE coordination?* If this question is not addressed, then researchers will continue to focus partially and narrowly on aspects of the problem of uncertainties. This is the theoretical challenge. If it is not addressed, then coordination practice informed by information processing will also remain limited in scope to just support for interactions. Consequently, typical sources of uncertainties such as constantly changing software requirements, communication delays and breakdowns, and knowledge transfer problems that emerge cannot be managed properly. This is the practical challenge. The question is addressed in Chapter 5.

In the *geography* perspective, it is assumed that concepts such as place, space, and physical distance and their relationships with ICT affect GSE coordination. In particular, physical distance bears directly on the resolution of coordination problems and it exacerbates these problems directly.[40] Marisa D'Mello and Sundeep Sahay also note that both the physical and social conceptions of place are far from being irrelevant in GSE.[41] However, explanations of how and why place and space exacerbate or resolve coordination problems are wanting in the literature. Space refers to the three-dimensional environmental structure of the universe offering opportunities for free movement of events, ideas, and people; place is a phenomenological locality invested with social meanings of appropriate behavior and cultural expectations.[42]

Since the dawn of the world wide web, spatial (of space) and placial (of place) issues in GSE have generated research interest on how to achieve coordination in the face of their constraints and opportunities. The theory and practice of virtual teamwork suggest that space gives much freedom for virtualization of activities and resources. However, this freedom is not unlimited because "places are woven together through space by movement and the network ties that produce places as changing constellations of human commitments, capacities, and strategies."[43] For this reason, research on GSE is characterized by scattered accounts of opportunities and limitations of both space and place. There are indeed practical and theoretical problems of GSE coordination due to geographical distance, dependency, delays, culture, and time. Against this backdrop, this book also attempts the following question to appraise this perspective: *how are place, space, and ICT implicated in GSE coordination?* The question is addressed in Chapter 6.

The works of James Herbsleb and his associates, as well as Erran Carmel and Ritu Agarwal, are the closest attempts to address this question. Interestingly, they could not address it satisfactorily because of the following characteristics about their works and findings. The problem with their analyses is that they do not present elaborate explanations of space and place in coordination. For instance, in their analysis of distance, dependencies, and delay, Herbsleb and his associates only describe how physical distance and delay affect dependencies between globally distributed team members. Similarly, Herbsleb and Grinter's examination of how physical distance interferes with effective communications to undermine architecture-, plan-, and process-based coordination largely describes the interfering role of physical distance and prescribes how it can be overcome. It does not explain clearly how space and place are implicated in coordination of task activities and information resources. Carmel and Agarwal's tactical approaches for alleviating physical distance to achieve coordination are also prescriptive, not explanatory.

Besides the exceptions, there is one group of GSE scholars that has researched the impact of physical distance on virtual teamwork in general but have left the particular problem of coordination in the background. For instance, Grinter and her associates refer to distance and interdependencies and, in fact, discuss collocation of task components as a coordination strategy. However, dependence, which is the key construct of coordination, is not treated explicitly, especially in terms of how it is affected by distance. Marcelo Cataldo and Sangeeth Nambiar rather studied the relationship between physical distribution and process maturity on software quality, but did not pay particular attention to coordination nor to the impact of place on it. Erran Carmel and Pamela Abbott's notion of "nearshoring" stresses that physical distance indeed matters, but coordination is marginalized in their explanations. The same limitation characterizes Pär Ågerfalk and his associates' framework that matches socio-cultural, geographical, and temporal distance to control, coordination, and communication.[44] In these studies, even though physical distance is acknowledged, it is perceived as a factor that engenders communication breakdowns, mutual knowledge problems, and socio-cultural bottlenecks.

Another group of scholars have studied GSE coordination but left the reality of physical distance in the background. For instance, Julia Kotlarsky and her associates advocate for a knowledge-based approach to coordination and discuss the importance of knowledge processes in coordination. However, their work does not explain the role of physical distribution in these processes. Similarly, Narayan Ramasubbu and his associates have discussed the relationship between spatial, temporal, and structural configurations of software project performance without paying particular attention to the placial and spatial dimensions of coordination.[45] A combined placial, spatial, and technological explanation is also missing in other explanations of coordination such as collective ownership and coding standards,[46] building shared meanings,[47] temporality,[48] expertise,[49] interpersonal trust,[50] architecture,[51] delegation,[52] evolution,[53] and sense and co-creation.[54] Consequently, there is a neglect of the complementary placial, spatial, and technological explanations of GSE coordination, and so knowledge of the relationship between geography and technology coordination remains scanty.

In the *organization* perspective, the assumption is that the global distribution of software resources and activities increases the proportion of uncertainty, openness, and indeterminacy in virtual teams. This increase is due to cultural and national diversity, erratic information exchanges, mutual knowledge problems, and politics. In GSE, the distribution produces direct effects such as task modularization and requisite variety in the resources and activities. Task modularization refers to the separation of software tasks and components into optimal clusters; and requisite variety means the desirable differences in a system match the differences in its environment. GSE organizations compensate for these direct effects with task standardization and team socialization. Task standardization refers to focusing attention on particular organizational control mechanisms to centralize them or make them uniform; and socialization manifests when team members develop clan relations and mutual understanding over long periods of engagement. The co-existence of direct effects and compensations points to a paradox of organization, understood as contradictory or conflicting yet interrelated factors that exist simultaneously.[55]

The paradox of organization has a bearing on our understanding of the nature of GSE coordination because it reflects in the simultaneous negative and positive relations between dependencies, uncertainties, and conflicts – known as coordination constructs. Negatively, the value of one of these constructs may increase as the other decreases when they are being managed to achieve coordination. For example, to reduce uncertainties between developers in different sites, organizations collocate software components. This increases collocated dependencies and, at the same time, reduces remote dependencies; time differences between developers (potential conflicts) are tolerated to ensure follow-the-sun development (smooth dependencies); and the practice of agile methods (potential uncertainties) is allowed alongside the practice of formal methods (certainty). Positively, uncertainties increase to undermine dependencies and conflicts increase to undermine uncertainties and dependencies. Likewise, intellectual and agile capacities are required of software developers who collaborate in the face of a virtual or synthetic reality, suggesting that management of uncertainties and conflicts between them goes together with the management of dependencies.

This paradox is currently acknowledged in research in organizational change and in virtual teamwork survival and improvisation. The understanding of virtual teamwork as a paradox is useful because it accounts for the remote distribution of resources and activities. It also provides implicit insights into how work is coordinated. But the paradox has not been applied explicitly or directly to explain the particular yet dominant concept of coordination in GSE. Currently, indirect application of the paradox to coordination characterizes existing explanations in which researchers have focused primarily and partially on one or another of the coordination constructs (see Appendix 1, Table 1).

As a result, the coordination constructs are currently quite isolated, leaving us with a parochial understanding of their relationships. The nature of coordination points to the quest for a conceptualization of its basic construct and integrated structure. Hence, this book also attempts the following question to appraise the organizational perspective: *how does ICT combine with organization for the achievement of GSE coordination*? To address this question, the paradox of organization is applied to interpret and reconceptualize GSE coordination in order to identify its basic constructs and structure. This is the focus of Chapter 7.

The four perspectives and instances pertaining to extant GSE coordination theory testify that researchers have addressed the problem of coordination in quite fragmented ways. The discussions of the technology, information, geography, and organization of coordination in extant research provide piecemeal accounts without adequate attempts at integrating them. Therefore, beside the appraisal of these four perspectives, there is the need for an integrated explanation of their interrelations in order to achieve adequate understanding. The absence of an integrated explanation is a significant gap in GSE coordination theory development, and one of the main reasons for these gaps is the absence of a virtuality approach to the development of coordination theory in existing publications. But if the aim is to develop a theory of coordination which explains the essential and enabling role of ICT, then we need to adopt a virtuality approach to integration.

The four perspectives reflect the assumption that the traditional organizational logics of rationality and indeterminacy are the only ones essential to GSE. Rational logic refers to the "extent to which a series of actions is organized in such a way as to lead to predetermined goals with maximum efficiency"; indeterminate logic refers to the idea that "interconnections among dependent parts are somewhat less constrained, allowing for flexibility of response."[56] This book argues that these are not the only logics that should inform our approach to the development of coordination theory. Following our premise that developing a theory of technology coordination requires an approach that prioritizes the role of technology ahead of information, geography, and organization, this book turns to virtual logic to take a different tack. Without sacrificing rational and indeterminate logics, this turn gives primacy to virtuality ahead of rationality and indeterminacy.

The new assumption premising this book is that virtuality is the primary logic of GSE organization – that virtuality is not merely contextual but essential to GSE coordination theory and its development. Virtuality reflects ICT-based representations and simulations of software development processes, spaces, and people that enable new coordination practices. Enabled by the combination of place, space, and ICT, virtualization of organization provides unlimited possibilities of organization and action that clearly transcend non-virtual teams. It also provides almost infinite opportunities for the mobility of people, information, tasks, and events.

For example, Arvind Malhotra and Ann Majchrzak report that "virtual team members rely on virtual workspace tools to coordinate knowledge that each individual brings to the team."[57] Their explanation of knowledge coordination amply demonstrates the epistemological (rather than mere contextual) importance of virtualization as an approach to the study of coordination. Julia Kotlarsky and her associates have also demonstrated the epistemological importance of virtualization by their explanation of how effective ICT-based connections enable knowledge coordination. However, their focus on knowledge, rather than technology coordination, is significant. Knowledge coordination, even in the virtual teamwork context, provides explanations of coordination that views ICT is an instrumental rather than an epistemological or essential issue. This assumption leads researchers to approach coordination in virtual teamwork with rationalization and indeterminacy as the only essential logics of organization.

The instrumental view of ICT derives from the assumption that virtuality is not an essential organizational logic. The virtuality of GSE organization is not assumed to be the central logic in these accounts, and so the technological, informational, geographical, and organizational explanations remain epiphenomenal to the core phenomenon of virtuality. By the neglect of a virtuality approach, thereby overlooking its essential logic, researchers of GSE coordination have been confined within the contradictory and traditional logics of rationality and indeterminacy. A theory of GSE coordination, which aims at the study of the roles of technology, information, geography, and organization, must progress beyond particular and epiphenomenal analyses toward integrated and essential analyses (Figure 2.1).

Theory and Philosophy

To initially appraise the four perspectives, several theories are drawn upon to offer us conceptual insights for explanations and predictions of how each of them is implicated in GSE coordination. These appraisals are made toward the eventual development of an integrated theory of GSE coordination through the overarching virtual approach. The choice of these theories is informed by their analytical roles in the appraisals and integration of the four perspectives. Each of these perspectives is itself an approach to the study of coordination. Therefore, each one also serves as a key criterion for judging whether or not any theory is an appropriate and a relevant analytical framework for each of the four research questions. For example, in the appraisal of the geography perspective on coordination, a theory with a geographical orientation is required for analysis of how places, spaces, and ICT are combined to achieve GSE coordination.

Appraisal of the technology perspective requires a theory with a technological orientation that has the potential to shed light on aspects of ICT that shape the management of dependencies within and across GSE sites. Such a theory must be characterized by technological realism. It must assume that technology is a real and generative mechanism in accordance with technology materiality[58] and critical realism.[59] Thus, in accordance with technology materiality, the assumptions of media ecology are used for appraising the technology perspective. The media ecology analytical framework assumes that "the medium is the message," meaning that the medium is primary and the message or content is secondary in the organization of human affairs. This assumption is drawn upon to explain how ICT media, rather than digital content, is primary in the organization of GSE. For example, according to media synchronicity theory, which draws upon media ecology, ICT media have the ability to "support synchronicity, a shared pattern of coordinated behavior among individuals as they work together."[60] This support is

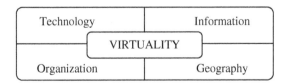

Figure 2.1   Integrative role of virtuality among the perspectives.

achieved through related communication processes called conveyance (transmission of new information) and convergence (discussion of individually pre-processed information). Media synchronicity theory also accords with critical realism because the idea of "capabilities of media" in the theory leads to an enriched understanding of the material ontology and causality of ICTs used for GSE coordination. Media synchronicity is a theoretical orientation that enables us to take a major step of overcoming the under-theorized role of technology in coordination. This study focuses on the technological (its material and realist ontological renditions) not because interpretive flexibility of ICT in coordination is irrelevant. This focus is because of the interest in exploring technological essence as an independent aspect of coordination.

To appraise the information perspective, this book draws upon the assumptions of Jay R. Galbraith's organizational information processing theory[61] and its adaptations by Michael Tushman and David Nadler,[62] by Richard Daft and Robert Lengel,[63] and by Richard Daft and Karl Weick.[64] GSE organization faces task uncertainties and equivocalities from within and without its boundaries. The theory assumes that an organization must design a structure that is capable of processing information in order to reduce uncertainties and equivocalities. It also assumes that different tasks create different information processing needs for the organization. The organization designs "mechanisms that permit coordinated action across large numbers of interdependent roles" in order to reduce the need and/or increase the capacity for information processing. The design strategy is either to reduce the need for information processing or to increase the capacity for it. This research considers GSE as a heavily uncertain task not only because of its complexity but also because of the task environment as well as inter- and intra-team dependencies across globally distributed sites. Therefore, information processing theory is relevant and appropriate for explaining how global software tasks are coordinated through management of information. Given that every organization is itself an interpreting system,[65] it is appropriate to adopt an interpretive approach to the study of how coordination is achieved by the use of GSE organizational design and information processing. Besides, information is itself a relative concept, as Albert Borgmann has indicated: it is a "sign: the fulcrum of the economy of information, and on it revolves the relation that mirrors the symmetry of humanity and reality, of intelligence and context, that undergirds every kind of epistemology."[66]

Appraisal of the geography perspective accounts for the roles of place and space in coordination. Given that GSE is an innovative activity, the choice of a geographically oriented theory must discuss other innovation inputs such as information, knowledge, and technology. This discussion is important because theories of the geography of innovation, especially in knowledge-based industries, have been questioned in the wake of globalization and the telecommunications revolution. Increased globalization after the fall of the Berlin Wall raised questions about comparative advantage in innovation which draws from location as a meaningful input.[67] In *The Death of Distance*, Frances Cairncross also questioned the geography of innovation with the argument that the internet had become the most powerful driver of innovation.[68] These questions suggest that information, knowledge, and technology are sufficient for innovative activities, especially in knowledge-based industries. But the quest for a geographically oriented theory in her book does not presume a linear model of innovation whereby scientific knowledge is the only source of ideas for new products for the market. Rather, theories of the geography of innovation assume that innovation results from multiple sources notably knowledge, location, technology, institutions, and experience. For instance, Maryann Feldman speaks of location in terms of "a geographical unit over which interaction and communication is facilitated, search intensity is increased, and task coordination is enhanced."[69] Location-based interaction and communication translate information into knowledge required for innovation through a non-linear process. The relationship between geography and the internet points to a general conception of space and a phenomenological conception of place. Space becomes general when combined with ICT because they constitute a virtual realm that offers opportunities for free movement of information. The phenomenological conception of place presents both relational and experiential understandings of it without sacrificing its material ontology.[70] Both conceptions, general and phenomenological, inform interpretations of how geography is implicated in GSE coordination.

Appraisal of the organizational perspective focuses on the paradoxical character of GSE and its coordination. The appraisal assumes that global distribution of resources opens up the organization to more environmental influences. As a result, the distributed software production system is characterized by indirect, less frequent, less important, and weak linkages between resources and activities. Thus, the software production system is characterized by an organizational paradox. For this reason, the quest for an understanding of how the character of organization is implicated in GSE coordination requires a theory with an organizational orientation that can handle the paradox in the software production system. Given this quest, rational organization theories are less suitable because of their sole strength in rational and unidirectional analysis. An organizational theory that has strengths in paradoxical analysis is most suitable for analysis. Wherefore, the analysis of how the organization is implicated in the coordination of GSE draws upon the paradoxical theory of loose coupling.[71] The theory of loose coupling has strengths in both rational and non-rational or indeterminate analyses. 'Loose' reflects the indirect, less frequent, less important, and weak linkages between organizational resources and activities which are at the same time "coupled" because they are characterized by various degrees of interdependencies. Loose coupling theory is the most suitable analytical framework for understanding the paradox of organization in terms of its nature as a system and its function as an adaptation strategy for coping with environmental influences in order to prevent their negative effects from spreading through the organization. The theory is also suitable for explaining coordination because its tenets have stronger correlations with the elements of organizational paradox – modularization, variation, standardization, and socialization.

For instance, modularization of loosely coupled organizational resources correlates with modularization of software components and teams. Variation among loosely coupled resources correlates with variety in locations, cultures, experiences, and times that characterize GSE settings. Since modularization and variation are direct effects of loose coupling which are compensated for by standardization and socialization, it is logical that they also have direct implications for coordination.

To eventually relate all these perspectives into an integrated theory of GSE coordination through the overarching virtuality approach, this book draws upon the dialectical method. The philosophical underpinning of this method is Georg W. F. Hegel's dialectical idealism which assumes a "world spirit" or idea – a system of abstract universals that explains countervailing things and events in nature and history.[72] The idea is neither spatial nor temporal; it is abstract, but finds its concrete expression in nature which is material. Therefore, absolute idealism interprets nature and history in terms of the primacy of the idea. The choice of dialectical philosophy and method to interpret how virtualization is implicated in GSE coordination is based on the assumption that digital logic (which is the essence of ICT) is the absolute idea of coordination. It is the absolute idea that finds its concrete expressions in coordination by plans and mutual adjustments that have long been upheld as the primary explanations of all extant coordination theories.

Chapter 3

Logic of Virtuality

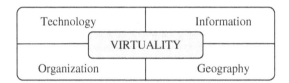

To virtualize a process (leading to information and communication technology [ICT]-based virtualization) is to remove physical interactions between people and/ or objects, and to apply ICT to the process. This is the summary of Eric Overby's argument undergirding his process virtualization theory.[1] The ability to virtualize a process with ICT depends on fewer needs for people to enjoy sensory experiences, for people to enjoy social or professional relationships, for activities in a process to occur concurrently, and for identification and control of people and their behaviors. When ICT can provide sufficient representation, reach, and monitoring capability in the process, then organizations will seek to achieve process virtualization.

Information systems and organizational studies owe a great deal to Eric Overby for his theory that explains and predicts process virtualization. His process virtualization theory remains a central explanatory and predictive tool for virtual teamwork, computer-based simulations, and ICT-based representations. However, a problem with Overby's theory is its assumption that, for process virtualization to occur, the dependent variables (that is, low sensory, relationship, synchronism, identification, and control requirements) must be opposed to representation, reach, and monitoring capabilities of ICT. The theory assumes that once technology can be used to simulate the sensory elements of physical reality, once technology can remove time and space barriers to enable process participation, and once technology can authenticate the people participating in the process and track their activities, then, given the absence of the dependent variables, organizations will seek to achieve processes virtualization. But there is evidence from information systems and organization studies research publications showing that high doses of sensory experiences, relationships, and identification are required among people and their activities, in addition to ICT capabilities at the same time, to enable the formation and function of virtual teams (Figure 3.1).

For example, the opposition between sensory requirements and virtualization in the theory is challenged by the need for face-to-face interactions between virtual team members, through traveling, to enhance process virtualization.[2] Face-to-face interactions have this enhancing capacity because they undergird subsequent relationships built through technology or they enhance existing ones which may be floundering. Similarly, the same face-to-face mode of interactions enables rapid development of shared identities among virtual team members.[3] Global software engineering (GSE) processes are typical examples of both ICT-enabled virtualization that are actually enriched or complemented, not opposed, by high doses of sensory experiences, relationships, and identification among globally distributed team members.

One of the main reasons for this problem is that the theory ascribes a moderating role to ICT, signifying its instrumental rather than essential status in process virtualization. Moderating variables are secondary to a theoretical model as they affect or alter the correlations between dependent and independent variables. The correlations, as well as dependent and independent variables, play the primary roles. In process virtualization theory, Eric Overby gives primacy to sensory experiences, relationships, synchronism, identification, and control, and so it is not surprising that he had to oppose them to process virtualization. With ICT in a secondary role, it is hard to see any complementation between these variables and process virtualization. Had he given primacy to ICT, he would have realized that its representation, reach, and monitoring capabilities actually require high doses of some of his independent variables to achieve process virtualization. He would have realized that both ICT capabilities and the independent variables (low sensory, relationship, synchronism, identification, and control requirements) complement each other on the one hand and complement process virtualization on the other.

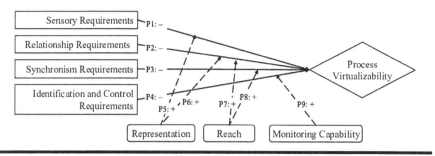

Figure 3.1 Process virtualization theory.

[*Source*: Overby (2008).]

ICT is itself characterized by loose coupling between its form and function on the one hand, and physical infrastructure on the other.[4] Computer software now allows for digital content to be self-referential. Digitization, representation, and simulation imply that new informational realities can be created out of information (not out of physical reality).[5] They imply an upward flexibility of information services freed from the fetters of underlying network infrastructure, as well as a downward flexibility whereby "a wide range of digital and physical networks potentially can provide the required interconnectivity and other functions."[6] Downward flexibility is the basis of integration of physical reality into ICT innovation.

ICT enables continuities or possible virtualizations of globally distributed processes, routines, and teams that can be generated by organizations by a huge magnitude. ICT innovations such as global telecommunication networks, digitization, and increased computational power have been extensively and intensively adopted by organizations. The lure of the virtual is so powerful that virtualization that is global in scope is the most dominant factor in organizational competitiveness and creativity.[7] The integration of place, space, and ICT in these processes is exemplified by concepts such as virtual worlds and virtual co-presence.

Organizational Logics

Logic refers to principles and methods for valid inference and demonstration. This book is not primarily concerned with philosophical logic which deals with questions of reference, identity, truth, ethics, aesthetics, existence, and necessity. It is rather primarily concerned with organizational logic which is the pursuit of understanding different principles and methods for valid inference and demonstration of different approaches to organization. This pursuit is helpful to clarify the traditional logics of rationality and indeterminacy as well as the proposed logic of virtuality.

Rationality logic assumes that the best way to conceptualize and manage an organization is to minimize losses and maximize profits or to produce the maximum output with the minimum input of resources. Economists and industrialists summarize this description as efficiency. Hence, organizational rationality is the "extent to which a series of actions is organized in such a way as to lead to predetermined goals with maximum efficiency."[8] Efficiency stands as the primary organizing principle of the rationality logic and translates into several methods – the initial one being specification of predetermined goals. But the method of goal specificity is only instrumental as an end of rationality. The more formative approaches are notable methods for goal implementation such as formalization, standardization, optimization, and centralization. In terms of ends–means inference and demonstration of the rational logic of organization, these methods validate rational actions in goal implementation even if the goals happen to be irrational.

Formalization of organization structure is the development of a cognitive and abstract mechanism that stipulates roles, role relations, and rules governing behavior. Formalized structures are symbolic, communicable, and modifiable means to enhance organizational efficiency. Formalization is the predominant method of Henri Fayol's administrative management approach to rationality. Herbert Simon's concept of bounded rationality in his theory of administrative behavior integrates formalization and goal specificity. Through formalization, organizations are able to achieve standardization and regulation of behavior, which are also essential preconditions for rational action. To standardize is to conform means and behavior to some agreed level of efficiency. Formalization and standardization in turn enable easy optimization of alternative actions for goal implementation. To optimize is to apply scientific methods to maximize efficiency and minimize inefficiency by systematic selection of behaviors. Optimization methods rely heavily on mathematical analytical techniques, and these techniques underpin Frederick Taylor's scientific management approach to rationality.

Indeterminacy logic assumes that organizations are more natural and open than rational. Following the biological meaning of the word organ (which connotes adaptation, selection, variation, and interdependency) this logic does not necessarily presuppose that organizations are deliberately constructed to pursue specific goals. Rather, it presupposes that organizations are collectives with complex goals and complex relations between normative and formative aspects of organizational behavior. Compared with the mechanistic system in the rational logic where interdependencies are constrained within rigid structures, the organic system in the indeterminate logic is characterized by loose coupling, flexible responses, equivocality, and uncertainty. Thus, rather than the search for optimal choice of action in rational organization, indeterminacy gives more priority to methods such as cooperation, motivation, and commitment within informal structures. Organizational theorists draw upon these descriptions to argue that effectiveness (ahead of efficiency) is the primary organizing principle of the indeterminacy logic. This principle promotes methods such as informality, cooperation, motivation, and institutionalization.

Without denying, but questioning, formal structures in organizations, indeterminacy focuses on the emergence of informal structures based on the personal characteristics of employees and their relations. For this reason, concrete and emergent processes of action characterize the behaviors that are used to achieve organizational effectiveness. Based on this reasoning, organizations promote cooperation and socialization among employees so that they integrate their individual knowledge and skills. This reasoning assumes that individual employees are limited cognitively and physically, and so conscious and purposeful means to induce their cooperation and socialization achieve greater effectiveness. Chester Barnard's conception of the organization as a cooperative system is informed by this assumption. Conscious and purposeful socialization requires employee motivation which Elton Mayo, Fritz Roethlisberger, and William Dickson describe in the "Hawthorne effect."[9] Out of loosely coupled activities in informal, cooperative, and sometimes unstable activities, institutionalized patterns emerge with time. Institutionalized patterns are stable and socially integrated activities which give an organization a special character and underscore its dynamic capabilities.

Informal, cooperative, motivational, and institutional methods in the indeterminacy logic are informed and validated by functional analysis (as compared with scientific analysis in the rational logic). According to Arthur L. Stinchcombe's definition of functional analysis, the consequences of employee behavior are also essential causes of that behavior.[10] Thus, behavior is explained by its function or consequence instead of some predetermined cause or goal. This is a cyclical form of reasoning, following Ludwig von Bertalanffy's general system theory, which emphasizes positive feedback and homeostasis (organizational stability, equilibrium, and survival).[11] Functional analysis identifies feedback loops in an organizational system which reinforces its causes or inputs.

Rationality and indeterminacy are longstanding logics of traditional organization. Moreover, rationality and indeterminacy are considered in organizational theory as essential logics but virtuality is not considered as such. Virtuality in organization theory rather functions as a quality or context of rationality or indeterminacy. It is quite indifferent to the essence of these two logics as exemplified in how ICT is perceived to play just an instrumental role in virtual organization – whether the organization is dominated by rationality or indeterminacy. Consequently, existing organization theory is quite deficient in terms of the development of a principle and related methods for valid inference which underscore virtuality as an essential and distinct organizational logic.

Coordination problems are ultimately traceable to dependencies, uncertainties, and conflicts, and efforts to address them manifest in notable conceptualizations such as differentiation and integration of resources and activities, arrangement of dependencies, information processing, organizational structuring, interorganizational relations, and distributed organizing. However, these conceptualizations are very generic and pitched at the high level of organizational strategy. At the management level, they can be further understood in terms of two predominant modes which reflect rational and indeterminate logics: mechanisms and processes.

On the one hand, coordination mechanisms are the reified, standardized, or crystallized versions of the processes, understood in terms of artifacts and protocols[12] and "pre-established plans, schedules, forecasts, formalized rules, policies and procedures, and standardized information and communication systems."[13] The use of this mode of coordination for organization is described by James March and Herbert Simon as "coordination by programming"[14] and by James D. Thompson as "coordination by plans."[15] Coordination mechanisms align with the logic of rationality because plans, standards, and programming signify the organization of resources and activities with predetermined goals for maximum efficiency. Hence, they align with the rational logic which organizing principle is efficiency.

On the other hand, coordination processes reflect non-physical and non-reified arrangements, behaviors, feedback, and mutual adjustments. These processes are used to handle dependencies among "shared resources," "producer-consumer relationships," "simultaneity constraints," and "task-subtask dependencies."[16] As organizing has increasingly become uncertain and equivocal because of task complexity, firm growth, and the environment, researchers have produced refined coordination processes such as energy in conversation,[17] social knowledge and procedural coordination,[18] relational coordination,[19] sense-making,[20] expertise coordination,[21] role structures and enactments,[22] team scaffolds,[23] and collaborative community.[24] These refined coordination processes show fewer constraints among interconnections between dependent entities, and allow for flexible responses. Hence, they align with the indeterminacy logic which organizing principle is effectiveness.

Virtual Logic

Coordination modes in GSE reflect the indifference of virtuality to these two logics because they are largely modeled after coordination theories of traditional organization wherein ICT is instrumental. Notable coordination mechanisms used in global virtual software teams are centralized software architectures and plans, standards, plans and formal mutual adjustments, transactive memory using standardized templates, process standardization, and technology platforms. Notable coordination processes used in global virtual software teams are team knowledge sharing and integration, ambidextrous coping, process agility, and modularization and information sharing. In these coordination modes, even where coordination is explained with ICT mechanisms, the explanation assumes that ICT is supportive but not essential to coordination, signifying that virtuality is not assumed in those explanations to be an essential organizational logic.

Virtuality as an essential logic of organization is first understood in terms of the peculiar task coordination challenges it addresses, just as the logics of rationality and indeterminacy are understood by their peculiar task coordination challenges they address. Richard Daft and Robert Lengel,[25] drawing from Charles Perrow,[26] explain organizational tasks by the relationship between their two main dimensions: variability and analyzability. Task variability refers to the amount and frequency of exceptional events in the task, while task analyzability refers to the amount of exceptional actions or behaviors and of time required to deal with the work exceptions. They match variability against analyzability to formulate two main descriptions of tasks: highly analyzable and lowly variable on the one hand; and vice versa on the other.

Highly analyzable and lowly variable tasks have certain and definite characteristics; and they are labeled as "problem tasks."[27] Problem tasks require rational measures such as mechanistic information processing, routine technology, and coordination mechanisms.[28] In software development, notable coordination mechanisms used are formal and analytical methods, plans, and architectures that specify requirements in systems design.[29] Conversely, highly variable and lowly analyzable tasks are characterized by greater degrees of uncertainty and equivocality;

they are labeled as "fuzzy tasks."[30] Fuzzy tasks require indeterminate measures such as organic information processing, non-routine technology, and coordination processes.[31] In software development, notable coordination processes are informal, flexible, and ad hoc communications alongside agile and experimental methods. In sum, the rationality logic aligns with problem tasks while the indeterminacy logic aligns with fuzzy tasks.

The GSE task has both problem and fuzzy characteristics; but it also has a distinct discontinuous character. Virtual organization in this context aims to span discontinuities between geography, time, organization, culture, work practices, and ICT.[32] For this reason, the GSE task can be distinctly characterized as discontinuous because of the global dispersion and remoteness (spatial and temporal) of its resources and activities. The new descriptions of global remoteness and dispersion add to existing descriptions of uncertainty and equivocality to give the task a subtle dimension (in addition to variable and analyzable dimensions). Plans and mutual adjustments are not sufficient to coordinate discontinuous tasks as the GSE literature suggests. In addition to coordination processes such as informal, ad hoc, and verbal communications, the literature suggests modes such as selection of developers from different locations and exploitation of their intellectual capacities and ambidextrous coping.[33] These modes are applied alongside virtual workspace tools for representation and simulation of software development activities that enable emergent coordination practices.

Second, virtuality as an essential logic of organization is understood in terms of the essence of ICT for coordinating the peculiar GSE task. ICT, which enables virtualization of software engineering, is also an instance of electrical media which, according to Marshall McLuhan and other media ecology scholars, stands in contrast with mechanical media. The typographical principles of uniformity, fragmentation, and linearity in mechanical media reflect rationalism, which is the central logic of the industrial revolution. Mechanical media was heavily influenced by Johannes Gutenberg's invention of the movable type and literacy as it dominated the industrial revolution. This domination occurred because mechanization reflects fragmentation of any process as in the assembly line whereby mechanization puts the fragmented bits in a slow sequence. One of the main reasons for slowness is that mechanical media is characterized by long spaces and time lags between action and reaction. Mechanical media does not easily provide total awareness of a production field to any manager because the media is unable to overcome long spaces and time lags. Therefore, any reaction to an action by a mechanical medium happens after a longer time period and/or at some distance from the action.[34]

Rather, electrical media provides high speed and total field awareness to managers of any production system. High speed bespeaks the accelerated pace of electronic information that overcomes space and time barriers. The acceleration also provides total awareness of a production field as physical commodities increasingly assume the character of information: "with electric technology all solid goods can be summoned to appear as solid commodities by means of information circuits set up in the organic patterns that we call "automation" and information retrieval."[35] Jannis Kallinikos calls this "work cognitivization" which he ascribes immediately to computer-based technology in organizations.[36] But the computerization is facilitated by a long historical process of developing and using various symbol systems for organizational resources and activities.

Virtualization of software engineering is a typical example of visual integration of space and time as well as production and consumption, all generated by ICT. The speed of technological information movement enables the constitution of virtual software teams whose members operate from different parts of the globe. This speed and coverage achieve total field awareness and organic interplay between activities and resources, pointing to the inherent coordination capacity of ICT. Virtualization of software development reflects total and organic interplay because it tends toward instant interrelations of the total field of resources and activities across space and time. Until the invention of electronic media, a virtual software development team was only an ideal condition in the mind. However, ICTs are extensions of the human central nervous system and they can be made to simulate human consciousness.

Those organic interrelations that the computer simulates by the instant electrical speed of information synchronization is a significant departure from other non-electronic media that moved things slowly. Non-electronic media is characterized by long lags between their actions and reaction. But the instant speed of electric information, and hence digital information, provides easy and instant awareness of reactions that allow for instant reaction. This instant awareness enables the design of interdependencies across space and time in global virtual teams. "Electricity unifies these fragments once more because its speed of operation requires a high degree of interdependence among all phases of any operation. It is this electric speed-up and interdependence that has ended the assembly line in industry."[37] Comparatively, those interdependencies managed with mechanical media, which preceded electrical media, were characterized by fragmentation.

The study of media is the study of environments because media is not just the conduit but the environment. For this reason, our study of ICT and its enablement of virtualization of software engineering is a study of the ICT environment. This is presumably an environment that is substantively different from the environment of mechanical media where rationalism is the dominant approach to management and research. Rationalism is an appropriate approach to study mechanical media environments because its mathematical and economical principles align perfectly with uniformity, fragmentation, and linearity. Conversely, electrical media is characterized by variety, integration, and dialectics, necessitating the proposed virtualization approach to the study of how ICT media is implicated in GSE coordination.

Media ecology scholars argue that "the medium is the message,"[38] not literally or mathematically as in the medium equals the message. Rather, they metaphorically imply a "complex, dialectical relationship between medium and message."[39]

The medium is primary, and the message or content is secondary in the organization of human affairs. This does not imply technology determinism because "we shape our tools and thereafter they shape us."[40] Our ICT media shapes us by its materiality whereby it enables and/or constrains without determining human behavior. Technologies enable or constrain in the sense that "they do not allow users to do whatever they want."[41] If ICT media is the messages in GSE, and if digital words generated from ICT media are also media and therefore also the messages therein, then GSE organization has a distinct media environment. It is a media environment because of the pervasiveness and preponderance of ICTs and digital information. ICT is right at the heart of software engineering, being the very task – it is both input and output. It is also the installed base that enables the composition of the globally distributed team as well as the majority of the interactions among team members.

The understanding of GSE based on media as environment is methodologically and analytically different from the understanding based on media as conduit. The study of ICT media with environmental methods implies approaching the study of technology as a medium which gives form to GSE coordination. This is an ecological approach, significantly different from the linear cause–effect approaches, reflecting the understanding of media as a conduit. Moreover, the study of ICT media with environmental analysis implies systematic and comprehensive unpacking of interrelations between ICT and geography, information, and organization. Methods and analysis of ICT in GSE as a media environment underscores the proposed logic of virtuality because virtualization enables the visual and managerial integration of space and time; and hence virtual analysis of coordination. Comparatively, methods and analysis of ICT in GSE overlooking media as environments reflect the logics of rationality and indeterminacy, where scientific and functional analysis of coordination leaves virtuality in the background as context.

The primacy of ICT media in virtualization and coordination is understood in terms of the idea that digitization of physical reality such as distance and global distribution of organizational resources creates their computer-based simulations and representations. "Virtuality occurs when digital representations stand for, and in some cases completely substitute for, the physical objects, processes, or people they represent."[42] As Albert Borgmann argues, representation is not just information *for* reality but information *as* reality.[43] Borgmann also calls this technological information to distinguish it from natural information (altars, foot trails, etc.) and cultural information (symbols, records, architectures, etc.). Technological information is not just a sign referring to a physical reality; it is also an icon or a simulation of both physical reality and other signs. Simulations show the power of virtualization to create new representations of distance and global distribution of organizational resources (Table 3.1).

By the organizing principle of creativity, the description of virtuality using simulated and represented physical reality to span global organizational discontinuities is qualitatively different from the description in collocated organizational settings.

Table 3.1   Organizational Logics: Principle, Methods, and Analysis

Logic	Principle	Methods	Analysis
Rationality	Efficiency	Specificity, formalization, standardization, centralization	Scientific
Indeterminacy	Effectiveness	Complexity, variation, informalization, cooperation, institutionalization	Functional
Virtuality	Creativity	Continuity, digitization, representation, simulation, disembodiment, location independence	Technological

Table 3.2   Organizational Logic and Task Types

Logic	Task	Description	Primary Software Development Methods
Rationality	Problem	Analyzable, certain, definite	Formal, analytical, specification
Indeterminacy	Fuzzy	Variable, uncertain, equivocal	Agile, prototypical, experimentation
Virtuality	Discontinuous	Subtle, remote, dispersed	Technological, intellectual, specification, experimentation

In collocated settings, virtualization is basically an issue of ICT-enabled interactions by mediation, augmentation, and support. Simulations and representations with ICT in collocated settings are primarily focused on products rather than teamwork or organization; that is, team members "operate on" representations.[44] When ICT-enablement in organization is limited to mediation, augmentation, and support, and when simulation and representation are limited to product design and learning, then virtuality is qualitatively insignificant and primarily instrumental. Comparatively, when ICT-generated indices of globally distributed people in a team are represented and simulated, and when team members "operate with" these representations to overcome problems of discontinuity, then virtuality is significant and essential to organization.[45] Thus, it is true that "virtual organizing is a matter of degree,"[46] but it is also true at the same time that virtuality in collocated organization is primarily instrumental and qualitatively different from virtuality in globally distributed organization where it is primary and essential.

Therefore, the GSE task can also be distinctly labeled as a "discontinuous task" that aligns with the virtual logic (see Table 3.2). Furthermore, if the organizing principle of rationality is efficiency, and that of indeterminacy is effectiveness, then the organizing principle of virtuality is creativity; that is, organizations' ability to create virtual spaces, global virtual teams, and emergent coordination practices to handle discontinuous tasks.

Chapter 4

Materiality of Technology

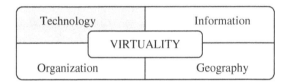

Technology	Information
VIRTUALITY	
Organization	Geography

Publication of *The Social Psychology of Telecommunications*[1] by John Short and his associates in 1976 brought forth a major explanation of remote team members' use of telecommunication technologies for electronic meetings. Their explanation was that remote team members' use of such technologies generated social psychological effects among them – labeled as social presence. Social presence refers to the technological representation of an individual and the degree of another person's awareness of the representation. Thus, the degree of social presence is the degree of representation and awareness of the interlocutor in an electronic meeting.

However, even without an electronic meeting, presence does not need to be taken literally. The social psychologist John Frederick Dashiell acknowledged this as far back as 1935, suggesting that presence is even obtainable when interlocutors know of each other's work on the same problem even if they operate from different rooms.[2] But Michael Tushman and David Nadler's work, which is far later than John Dashiell's, takes the literal view of presence, placing emphasis on verbal, ad hoc, and spontaneous communications that are generated by the face-to-face mode.[3] The difference between the literal and non-literal views of social presence is that Tushman and Nadler's view was formulated within the context of research and development (R&D) work. R&D is a particular type of work that falls within Charles Perrow's complex or non-routine work category, where tasks are difficult to analyze and have many exceptions.[4] The literal view is also acknowledged by Richard Daft and Robert

Lengel in their media richness theory, where they re-define non-routine tasks in terms of a combination of high uncertainty and high equivocality.[5]

Because the literal view is confined to non-routine work which is just one of Perrow's four work typologies, it is possible that routine work for instance – which is analyzable and has few exceptions – may align with the non-literal view of social presence. This alignment is possible since, as John Dashiell says, presence is obtainable when interlocutors know of each other's work on the same problem even if they operate from different rooms. But it is even possible that non-routine work can justify the non-literal view of presence when the assumptions of Daft and Lengel's media richness theory are taken into account. Their theory assumes that the basis for organizational information management is the clarity provided by media type not necessarily by data. Its dynamics are that rich and personal media such as group meetings provide more clarity and are suitable for non-routine work, and less rich and impersonal media such as computer databases and telephones provide less clarity and are suitable for routine tasks. Given that the theory hinges on media rather than data, it is possible that the enrichment of computerized media with more visualization, virtualization, simulation, and three-dimensional representations may justify their suitability for non-routine tasks. Both theories of media richness and social presence provide room for their extension with information and communication technology (ICT) innovations, as witnessed in research in computer-supported cooperative work (CSCW).

Terry Winograd and Fernando Flores spearheaded the early developments in CSCW through their explanation of the relationship between computer technology and the design of cooperative work systems. The basic assumption in their theory of design, technology, and action is that language is action. Based on human–computer interaction (HCI) design of technology that represents language, a large range of communication and coordination actions are made possible. The design of ICT systems that enable remote workers to observe their own actions in linguistic terms define the idea of representation.[6] By computerized representation in a system called the Coordinator (a conversational system), Winograd and Flores demonstrate that remote workers are further enabled to design their actions together, recognize breakdowns, and respond to them.[7] Their cooperative design of actions and handling of breakdowns are generated by requests, promises, assertions, and declarations which are fundamental linguistic actions according to speech acts theory.[8] Winograd and Flores' language-as-action perspective points to a language/action ontology of ICT that acknowledges a non-literal view of social presence and possible enrichment of media to make them suitable for non-routine work.

Essentially, the theories of media richness and CSCW are extensions of John Short and colleagues' non-literal view of social presence theory. Because of the non-literal view, Winograd and Flores' work constitutes a critique of Tushman and Nadler's as well as Daft and Lengel's literal view of social presence. The heart of their critique and hence contribution is the possible enrichment of technological media to increase social presence among remote workers, even if it is impossible to achieve full social presence through a technological medium. Possible enrichments have indeed been witnessed in the process of continuous ICT innovation over the past half-century. The ICT innovation process is basically generated by the combination or convergence of telecommunication and computation technologies. John Short and colleagues' formulation of social presence theory was largely based on telecommunication technology, and quite devoid of computation technology. CSCW formulations rather include telecommunication and computation technologies, and were the kernels for the development of group-decision support systems (GSSs) that are more advanced than the Coordinator.

Alan Dennis and his associates' media synchronicity theory[9] is the most cogent explanation of GSS development. GSSs are highly enriched with powerful telecommunication mechanisms and digital innovation of computer hardware and software. They enable social presence in non-routine work. Synchronicity exists among remote workers in a virtual team when they exhibit a shared pattern of coordinated work behavior with a common focus; media synchronicity is "the extent to which the capabilities of a communication medium enable individuals to achieve synchronicity."[10] Media synchronicity theory assumes that the fitness between media capabilities (including richness and synchronicity) and communication needs of work shape media appropriation and use. Based on the digitization of information and technologies, and their combinations, communication with GSS is understood in terms of two main processes – conveyance of information and convergence of meaning. Conveyance refers to electronic transmission of large amounts of data and subsequent retrospective analysis; convergence refers to electronic transmission of high-level abstractions of information that enable shared meaning among virtual team members. Remote work coordination (a process of convergence of meaning) has greater need for media synchronicity than conveyance processes.

GSS enable social presence in non-routine work such as "electronic brainstorming"[11] where remote members exchange ideas in an almost simultaneous manner. By the power of telecommunication and computation technologies, virtual teams using GSS are argued to even outperform collocated groups interacting verbally.[12] But there are also reports on negative performance outcomes from virtual teams supported by these systems.[13] The mixed performance outcomes signify that GSS tools do not guarantee certain support for virtual teams; nor do the outcomes signify that GSS are neutral or indifferent tools. According to the principle and reality of media enrichment due to continuous ICT innovation, they are neither neutral nor indifferent. The premise for the non-neutral nature of ICT is media ecology theory which assumes that "the medium is the message."[14] A GSS medium is the message because it embeds itself in the message which it conveys to generate a symbolic relationship with the message, sender, and receiver. When GSS media are used by virtual teams for remote communications, the media embed themselves in the messages, which exchange it enables. They embed with their repertoire of multiple formats and channels in messages to influence how receivers perceive those messages. Thus, the medium, message, communicator, and receiver generate not only a symbolic relationship, but also a functional one.

The symbolic and functional relationships between GSS media, messages, communicators, and receivers in global software engineering (GSE) are indicative of an electronic media environment which must be subjected to media ecology analysis. Media ecology is the study of media as environments.[15] Media environments in GSE refer to the range of possible coordination actions that GSS technologies enable or disable. This is not technology determinism but rather technology materiality. Determinism assumes a forceful and coercive role of technology in human affairs, while materiality assumes its enabling or disabling role.[16] Therefore, the symbolic and functional relationships among media, messages, communicators, and receivers in the GSE environment are explained in terms of the materiality of GSS media therein. Media ecology is the preferred analytical framework for explaining the materiality of GSS media in GSE coordination because it embeds the theories of social presence, media richness, and media synchronicity. All these embedded theories assume that ICT media are material, that they constitute environments, and that they generate psychological effects in users of the media.

Note that the conjunction of ICT materiality and relationships among media, messages, communicators, and receivers does not mean that sociomateriality is the framework for analyzing the materiality of GSS media in GSE. Sociomateriality, informed strongly by assumptions of Anthony Giddens' structurational theory,[17] essentially denies the existence of ICT materiality and social phenomena such as institutions, norms, and discourses as separate entities. Rather, sociomateriality lays analytical emphasis on emergent relationships at the intersection between materiality and social phenomena.[18] This relational ontology among ICT materiality and social phenomena is expressed by Wanda J. Orlikowski and Susan V. Scott as "constitutively entangled."[19] Their distinctions between materiality and society are only analytical: that is, ontologically, they do not have separate existences but their sociomaterialities are in various ways constituted by their relationships. But ICT media have physical and digital materiality that can be arranged into various forms to endure across different places and times.[20] It is true that ICT media are inseparable from social agency and implications in terms of development and function. Yet it is also true that once they are developed as particular objects they are separable in terms of materiality and structure. They are separable from different places and times and they also acquire new material and functional identities across different places and times. At any new place and time, an ICT medium existed before it was brought into that position or relation, and the medium may continue to exist beyond that place and time.

Consequently, we must also prefer the materiality of ICT media to their sociomateriality. The analytical framework of sociomateriality and other similar ones underplay ontological dimensions and rather emphasize social constructivism and interpretive flexibility. The basis of this preference is that materiality of ICT media is the structural basis of virtuality. ICT media have material structures that enable virtual integration of space and time as well as production and consumption of software in GSE. This is because software is an instance of electrical media which, according to the assumptions of media ecology,[21] differ structurally and materially from mechanical media. Instant electrical speed of information synchronization generated by ICT media is a significant departure from mechanical media that process things relatively slowly. The transformation of software development, a social activity, from collocated to virtual organization, mainly by ICT media, is testimony to the material ontology of this electrical medium. Its totalizing effects on organization, understood by how it provides total and integrated awareness of a production field, point to its distinctive materiality.

Even a closer examination of Eric Overby's process virtualization theory validates the claim that the structural basis of process virtualization is the materiality of ICT media. The theory's main assumption is that when ICT can provide sufficient representation, reach, and monitoring capability in the process, then organizations will seek to achieve process virtualization. The theory underscores the material structure of ICT media while eliminating people's sensory experiences, social or professional relationships, and identification and control of their behaviors. Thus, it emphasizes the existence of ICT media as separate material objects, which become functional for process virtualization when they are brought into an organization's space and time. It does not ascribe only a functional and relational ontology to ICT media; rather, it traces their functional and relational ontology to their structural and material ontology.

We must understand GSSs, which are instances of ICT media in GSE, in terms of their digital materiality. A GSS application is largely software but it is no less material than any other physical artifact because it has matter (has weight and occupies space), it has practical instantiation across different places and times, and it has significance especially as technology-in-practice.[22] Philip Faulkner and Jochen Runde[23] truly refer to computer software as non-material, but their reference must be understood in context. In their research, the context of use was to distinguish software (non-material) from hardware (material) and yet to argue that indeed non-material software is a technological object that is critically real. Software is critically real because it has material structure that lies at the heart of all computer-based information systems. To argue, with such critical realism, for the materiality of ICT media in GSE is to argue that is a generative mechanism that lies in the real domain – generating actual events that are empirically observable.[24] This means that it has separate existence as a structure, it can generate different functions across different places and times, and yet maintain its structure.

How and why then is the materiality of GSS media implicated in GSE coordination? Given that the study of media is the study of environments, we need to understand the materiality of GSS media in the context of how they generate an environment of interrelations between the media, messages, communicators, receivers, and software engineering. Such an ecological analysis of the GSE environment helps to explain how and why coordination is achieved by the materiality of technology media.

Coordination is the managing of dependencies between activities. This is the generic definition which seems to imply that the understanding of coordination is regardless of the type of dependencies. However, given that GSE implies globally distributed and collocated software engineering (since the former is a derivative of the latter), we have to identify two key types of dependencies: those that reflect global distribution and those that reflect collocation. Those that reflect distribution are dependencies mediated by factors such as physical space, time, and socio-culture; and those that reflect collocation are mediated by greater magnitudes of these factors including technology.

Based on the global description and structure of GSE, those that reflect distribution portray what I call global dependencies. The GSE literature problematizes such global dependencies by emphasizing how technology, space, time, and socio-culture generate uncertainties and conflicts that disturb the dependencies.[25] The problematization of global dependencies is due to how difficult it is for mechanical media to resolve them. Mechanical media, by their materiality, are characterized by low speed, space, and time lags between action and reaction, repetition, uniformity, fragmentation, and linearity. Examples are photocopiers, printers, cars, books, pens, and buildings. Their weaknesses in the face of managing dependencies that are mediated by global distribution of software engineering have continued to pose GSE coordination problems. This is because mechanical media, as well as space, time, and socio-culture, remain real and important aspects of the GSE environment. Yet the practice of GSE continues unabated, and organizations are continually achieving coordination across the world in spite of these problems.

The explanation of how and why the practice is continuing amid coordination successes is that it is relatively easier for electrical media such as the internet and GSS to resolve the problematized global dependencies. Besides the relative ease, the resolved dependencies mimic or orientate toward collocated software engineering because the virtualization that leads to GSE is intended to create a simulated collocated environment. The collocated description and structure of these dependencies suggest that they portray a different character which I call village dependencies (borrowing from Marshall McLuhan's epithet, "global village"). In view of the relative ease by which electrical media resolves village dependencies, they are not problematized in GSE research as much as global dependencies are problematized. But that should be expected because of two main reasons. First, electrical media are characterized by high speed, and, because of their materiality, they generate space and time synchronicity between action and reaction. Electrical media also generates total field awareness; integration of people, and task and information, as well as dialectics. Television, radio, the internet, and GSS are typical examples. And second, the dominant inputs and output of software engineering is digital information which is also electronic.

To wit, village dependencies and electrical media dominate the GSE environment through their materiality. Wherefore, the relative ease by which electrical media enables the management of village dependencies overshadows global dependencies and their management with mechanical media. It is therefore appropriate to interpret the GSE environment as an instance or microcosm of Marshall

McLuhan's global village. Global is the adjective; and so, it is secondary to village which is the noun and, hence, primary. By this media ecological interpretation, it is right to ascribe the continued coordination GSE success across the world to the materiality of the internet and GSS. Here, the virtual logic of electrical media trumps the rational logic of mechanical media in the explanation of GSE coordination. Virtual logic aligns with village which is the solution; rational logic aligns with global which is the problem.

What about the software engineers who are key components in the GSE environment? How does the materiality of the internet and GSS media in this environment shape their coordination actions? One of the core assumptions of media ecology is that media are environments because they generate ecological rather than additive changes. Marshall McLuhan, for instance, argued that people adapt to their media environments by adjustments of their sense ratios. Hence, the primary medium of an age develops or sharpens one sense or more, thereby affecting the person's perception. In 1998, Neil Postman gave an address at Denver, Colorado, entitled "Five Things We Need to Know about Technological Change." In this address, he also argued that perceptual change is induced by technological media. In *Technopoly*, he had earlier called attention to the fact that every technology contains a powerful idea that changes the sense and perception of the user:

> To a man with a hammer, everything looks like a nail. ... To a person with a pencil, everything looks like a sentence. To a person with a TV camera, everything looks like an image. To a person with a computer, everything looks like data.[26]

It is understandable that the GSE environment is an environment of computer technology media because of two main related reasons. First, computer technology media have changed the coordination structure of software engineering by transforming conventional or global dependencies that characterized collocated settings into village dependencies that characterize technology-mediated settings. The numerous ICTs have come to constitute a material stage on which software engineering is undertaken. Hence, they have come to constitute an environment which does not necessarily determine software engineers' actions but has enabled redefinition of their range of possible actions. In this media environment, global dependencies have been reduced to village dependencies.

Second, if the range of a global software engineer's actions has been enlarged or reduced by computer technology media, then coordination success presumes that he has adapted to this environment. In this adaptation, he does not only learn about how to use ICTs to work; he also modifies them to address typical GSE problems such as mutual misunderstanding, eleventh-hour changing customer requirements, and socio-cultural diversity. ICTs in GSE sharpen software engineers' senses of hearing and seeing digital words which make the GSE village better in many ways than the traditional village. Conversely, software engineers reprogram their ICTs to further enhance their capacities to hear, see, and act with their bodies and minds to address the typical GSE problems. Therefore, village dependencies underscore the non-literal view of social presence among engineers.

Neil Postman is right to say that "[t]o a person with a computer, everything looks like data." Digital data are the most dominant resources both in the mind of the global software engineer and in his environment. Since digital data highly engage the visual and aural senses of software engineers, both their oral and written sensibilities are aroused and exploited to address their typical problems. Village dependencies are first created in the minds of global software engineers before they are externalized in practical coordination actions. Digital words and images reflect electrical media while printed words and images reflect mechanical media. Thus, there is an interiorization of digital words and images by global software engineers in their work. Digitalization has the totalizing effect of field awareness through speed, synchronicity, and richness which stimulates the mind with what McLuhan calls "cosmic consciousness." Cosmic consciousness among global software engineers is their continuous experiences of closeness and interdependence and social presence, which underscores the village metaphor.

Therefore, in spite of the symbolic and functional relationships between the GSS media, message, communicator, and receiver, the materiality of the media becomes the focus of understanding why Dennis and his associates argue for the flexibility of GSS in their media synchronicity theory. In short, while relationship is the focus of analysis, the medium is still the message; the richer the medium, the higher its synchronicity for convergence processes, and the greater its propensity to coordinate work among global virtual team members.

Materiality of GSS and GSE Coordination

Software engineering is an instance of research and development (R&D), requiring high degrees of spontaneous, collective, informal, and, sometimes, synchronous interactions to get the work done. For this reason, using ICTs to enhance these interactions is crucial for successful coordination. In the context of global distribution, ICTs do not merely support such interactions; they actually enable interactions among remote developers. Without this enabling, optimal information sharing across sites would be impossible. Digital information that is captured, processed, stored, secured, and transmitted by ICTs is both a tangible and omnipresent resource for GSE. It manifests in representations such as text, voice, and images, and their generation, processing, storage, and sharing with ICTs characterize the majority of developers' actions. Although it is well known that software code development is done mostly in isolation because of the amount of concentration needed by the developer, it is also well known that developers work in teams, signifying a high magnitude of task dependencies between them.[27] Thus, GSSs, which facilitate information generation, processing, and sharing across sites, are reported in the virtual teamwork literature to be crucial for information generation, processing, and sharing.[28] GSSs make a significant contribution toward the management of cross-site dependencies and uncertainties.

The materiality of GSS is evidenced by four main functions of the technology for coordination which are espoused in existing publications on electronic meeting systems used in both general virtual teamwork contexts: mutual understanding, new assignments allocations, learning, and agility.[29]

First, in view of the task complexity and its collaborative imperative, GSSs contribute significantly to mutual understanding by reducing uncertainties across sites. Through teleconferencing, scattered information is shared and understood as required for team cohesion and collective decision-making. Verbal, textual, and graphical pieces of information are shared by virtual team members to enhance their mutual understanding. Second, GSSs serve as platforms for new task allocations in the face of eleventh-hour changes in requirements that generate increased uncertainties. In these situations, project managers verify previously assigned tasks during teleconferences and allocate new ones. Third, GSSs enhance learning among distributed team members. Although developers seek expertise from more experienced developers through other communication media apart from teleconferences, they also learn from remote colleagues' reports that are shared during teleconferences. Fourth, GSSs serve as important antecedents for agile development that is more necessary for dealing with eleventh-hour requirements. They are important antecedents because meetings pave the way for individual developers to work on the changes subsequently. Without antecedent teleconferences, team members only rely on emailing for collaboration, and that usually encumbers their agility considerably because emailing is asynchronous.

The materiality of GSS is evidenced by two additional functions which are distinctive to the GSE context, and which underpin this book's contribution of a technological explanation of coordination. These functions are ready information access and multitasking.

GSSs contribute to the management of dependencies and uncertainties by enabling information sharing when developers participate in electronic meetings from their own desks. From their desks, and hence from their computers, they have ready access to all information that needs to be shared even if such information was not predetermined to be shared at the beginning or before a meeting. In collocated contexts, software engineers usually leave their desks and converge at a meeting room with only predetermined information to share. This means that, in case there is an emergent need to share non-predetermined information during the meeting, the developers would either postpone the sharing or have to excuse the others to fetch it from their desks. This can delay or undermine decision-making. Teleconferences, therefore, ensure every developer's ready availability and access to all their information on their computers.

Multitasking during teleconferences can undermine the management of dependencies because an engineer's low concentration or partial listening to meeting discussions can cause him or her to miss vital information from the exchanges. The technology could show on a screen that an engineer is active or participating, but he or she may be working normally and may not even be concentrating on what is being discussed in the meeting. The participant may be wearing headphones with sound, but may not be listening or may only be partially listening. At the same time, there are significant coordination benefits generated by multitasking during electronic meetings. During such meetings, engineers do other important things apart from participating in the meeting, and only switch back on when the meeting convener specifically calls upon them to explain something. This multitasking and less-concentrated mode of participation allow them to continue their work in meetings where their participation and contributions were very marginal. The modes of participation-at-the-desk and multitasking explain how, through electronic meetings, new external requests can be received and addressed in the course of a meeting.

Each of the six coordination functions of GSS discussed above addresses one of three factors of dependencies that characterize GSE: those related to task characteristics, those generated by group characteristics, and those related to the global distribution of resources. The GSS serves as a platform for mutual understanding and new task allocations help in managing dependencies related to the complexity of software development. The function as a platform for learning and a precursor for agile development helps in managing dependencies engendered by experiential differences among distributed team members. The function as a resource for ready access to information and for multitasking helps in managing interdependencies related to the global distribution of resources (see Table 4.1).

These three factors of dependencies (software engineering complexity, experiential differences, global distribution) combined with GSS are interesting because they constitute instances of the broad constructs that Alan Dennis and colleagues[30] espouse for the analysis of electronic meetings – task, group, context, and technology. However, these factors are peculiar to GSE; they are not directly reflected in the

Table 4.1   Relationships between GSS and Management of Dependencies

Functions of GSS	Related Dependencies Managed	Dennis et al.'s Constructs
Platform for mutual understanding	Software development complexity	Task
Platform for new task allocations		
Platform for learning	Experiential differences	Group
Precursor for agile development		
Resource for ready availability and access to information	Global distribution of resources	Context
Resource for multitasking		

constructs of the research model of Dennis and colleagues, and they represent a different context altogether. For example, the combination of the global distribution of resources and teleconference technology is very foreign to their model. But its broad constructs are applicable to both sets of instances. Therefore, it serves as a useful framework for explaining the materiality of GSS in GSE.

The bases of the two distinctive coordination functions of GSS and how they contribute toward a technological explanation of coordination need further explanation. Both functions result from the inseparable combination of GSS and global distribution or distance. Without any one of these bases, these roles may not manifest, and other roles such as mutual awareness and timely information sharing may be limited. Thus, it is interesting to note how the electronic meeting, which represents one way of addressing coordination challenges brought by distance, also relates with the self-same distance to address those challenges. To wit, the self-same global distribution can engender coordination challenges and opportunities, and it is through the understanding of the materiality of GSS that one can make a distinction between the challenges and opportunities. Thus, GSSs are not to be deemed as a mere reactive measure to address coordination challenges. Rather, they should be deemed as proactive measures to actually enhance coordination by facilitating information generation, processing, and sharing by engineers across sites.

When the materiality of GSS enables drawing of coordination benefits from distance, we should understand them as resources of exploiting spatial or geographical structures. This understanding is a challenge to previous perceptions of the globally distributed structure of GSE. Previously, this structure was largely deemed as a negative factor, and rightly so when it connects with problems of culture, excessive communication delays, and mutual knowledge sharing. Even when it is perceived as a positive factor, it is because of benefits such as closer-to-market development and associated reduced cost, access to a global pool of technical and experienced developers for innovation, and continuous, 24-hour or follow-the-sun development. Moreover, previous research views GSSs only in terms of their reactive capacities to address distance-related coordination challenges. The materiality of a GSS suggests, however, that global distribution is indeed a coordination opportunity. It is an opportunity because it combines with a GSS to enable coordination.

Enablement of coordination due to the combination of global distribution and GSS is proven further from the fact that they induce developers' flexibility-in-participation or flexibility in their communicative behaviors. The manifestation of flexible communicative behaviors indicates that the GSS media have created an environment of reduced structure overload. Structure overload is a situation where the process of information generation and processing is constrained by the imposition of structural and processual rules and regulations.[31] These constraints are normally experienced by organizational units attempting to adapt to environmental demands for flexibility yet frustrated by rigid organizational structures. Thus, the imperative for face-to-face meetings to be held in special meeting rooms subsumes structure overload because it constrains participants' full access to their computers for information generation,

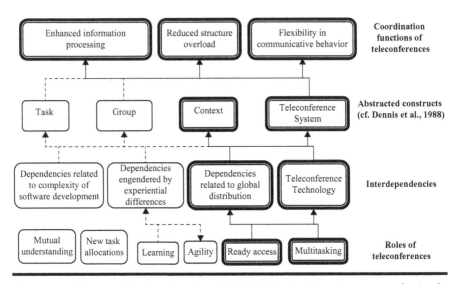

Figure 4.1   The coordination functions of GSS (distinctive functions are emphasized).

processing, and sharing. But GSS enables participants' full access to their information, especially for managing dependencies (see Figure 4.1). Thus, flexible communicative behaviors and reduced structure overload are one set of processes that explain the materiality of technology coordination in GSE coordination.

When GSSs function to induce flexible communicative behaviors, they are understood as occasions for weakening meeting structures – that is, occasions for de-structuring. This resonates with Stephen Barley's idea that technology is an occasion for structuring the social and institutional orders of organizations.[32] But the materiality of technology approach to coordination points to the need to observe more closely the object and influence of the structuring occasioned by GSSs. The object is the structure of the meeting genre which has its own substantive conventions.[33] This object is influenced by GSS use to result in organic information processing. The contrary influence by collocated meeting renders the genre more structured, induces more rigidity in participants' communicative behaviors, and does not reflect organic information processing.

In software engineering and R&D literatures, there is talk about the necessity of organic information management through face-to-face, informal, and spontaneous communications to coordinate work. But through global distribution, organic information management is augmented with mechanical information management through technology-mediated communications and formal organizational structures. This augmentation undermines information management in totality because more than less organic information management is required for coordinating a complex R&D undertaking like GSE. However, the materiality of GSS suggests that organic information processing may not be undermined if the use of it occasions de-structuring of the conventional meeting genre to enable multitasking and avail ready information. From the perspective of their form, GSSs are significantly different from face-to-face, informal, and spontaneous communications. But from the perspective of their functions, they enable developers' desire for flexible communicative behaviors and reduced structure overload in GSE.

Therefore, we should also not understand information management (mechanic or organic) with GSS for coordination just in terms of technology characteristics. Rather, we should understand it in terms of its structuring influence. The de-structuring of the meeting genre suggests that information management is a more elastic concept involving technology, structure, and task considerations. Hence, the degree of de-structuring will determine the degree of organic information processing, and vice versa. By this reasoning, multitasking, ready access to all information, flexible communicative behaviors, and reduced structure overload can be understood together as unique substantive conventions of the de-structured meeting genre. A unique substantive convention of a genre refers to its material or function or service for a particular purpose as opposed to its mere structure.[34] These substantive conventions of the de-structured meeting for organic information processing are unique because no other communication technology engenders such processes.

In sum, the bases and consequences of these two distinctive coordination functions underscore the idea that GSSs do not only draw positive benefits from distance but also induce flexible communicative behaviors among engineers. The de-structuring influence of GSSs implies that they can be deemed as a facilitator of organic information management, functionally speaking. But this functionality should be understood in the context of the software engineering task because the nature of the task demands organic information management. Without this demand, as in the nature of other non-R&D tasks, this functionality will neither be desired nor be realized. This implies further that we should understand the coordination function of GSS in terms of the materiality of the technology and its relationship with the complexity of the software development task and global distribution. It is only by this combination that GSS as an occasion for de-structuring can be seen as a sensible coordination process.

Project managers can learn from these explanations of how and why reduced structure overload and flexibility in communicative behavior are essential and strategic rather than normative conventions. Having proven to be distinctive and substantive to the de-structured meeting genre and for organic information processing therein, managers can draw upon these conventions to improve upon coordination processes in GSE projects. The conventions can also serve as a backdrop against which managers can evaluate requirements of GSS design for GSE coordination. In this evaluation, GSS facilities that occasion de-structuring should be considered as crucial requirements for organic information management.

Chapter 5

Management of Information

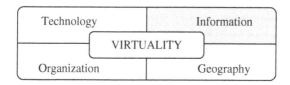

Information management through generation, processing, and sharing is critically required in global software engineering (GSE) primarily because of the problem of high degrees of uncertainties. Distance, task, and interaction technologies are additional sources of uncertainties in these work configurations that combine to challenge traditional information management requirements for coordination. High uncertainties predominate this work configuration as a persistent problem because they are emergent and varied. Emergence and variation of uncertainties challenge the stability and certainty that software organizations need to survive. At the same time, emergence and variation of uncertainties induce software engineers' dexterity and innovation to make their organizations responsive and resilient. Thus, the issue of high uncertainties is a practical and theoretical puzzle in information management terms. Given the logic of virtuality which provides visual integration of space and time as well as production and consumption through information and communication technology (ICT), what information needs to be generated, processed, and shared? What technological and human resources are needed to generate, process, and share information? And how and why should the technological and human resources be mobilized and utilized to generate, process, and share information? These are longstanding questions for both managers and researchers who are interested in the coordination of GSE. This chapter seeks to address these questions in order to proffer knowledge about how and why the problem of emergent and varied uncertainties are addressed through information management for the achievement of coordination.

The preliminary step in this endeavor is the relational conceptualization of information (instead of an individual conceptualization). Information is a relational concept because it is a sign that is signified by its mediation between a person and reality. This is Albert Borgmann's[1] philosophical theory of information which is not contrary to the scientific theory of information espoused by Claude E. Shannon and Warren Weaver.[2] For in their mathematical theory of communication, information is explained not as an isolated entity but as a relation among phenomena such as source, transmitter, channel, and receiver during a communication session. However, these relations are implicit in the mathematical model which is strong in simplicity but weak in its consideration of context. In Borgmann's philosophical rendition, information as a relational concept is explicit. Its fundamental structure is this: "INTELLIGENCE provided, a PERSON is informed by a SIGN about some THING within a certain CONTEXT [emphases in the original]."[3] The reference function of the sign is strongly rendered here like a mirror showing humanity and reality in a symmetrical relationship.

Borgmann describes the information relation in terms of an evolution from nature through culture to digital technology (as different from other technologies such as the motor). The creative and integrative principles of virtuality logic are visible in his idea of "information *as* reality" – that is, reality created by digital technologies in this third evolutionary stage. In the first evolutionary stage, the idea of "information *about* reality" corresponds to the natural description whereby the world of nature is made perspicuous with records and reports. In the second stage, the idea of "information *for* reality" corresponds to the cultural description whereby the world is made more prosperous with recipes, plans, theories, and constitutions. In the technological description, information *as* reality refers to the "hope that a truly unencumbered and clearly structured *possibility space* would engender a burst of actual creativity [emphases added]."[4]

The binary digit, which operates with a base-two logarithmic function, is the core principle of the computer (digital information technology). This is the mathematical function undergirding Shannon and Weaver's scientific model of information. This function does not by itself enhance information about reality, because the richness of reality does not make it possible to represent everything in bits. Yet it provides the structured possibility space for information bits (zero, one – digital signs) to be combined infinitely. Until the invention of digital information technologies (computer screen, storage, transmission capacity), the possibility spaces were limited and circumscribed by the octaves and keys in musical instruments and other grids such as graphs and maps. However, digital information technology, through its thousands of screen pixels and hundreds of colors each pixel can have plus increasing bandwidth and processing capacity, has enlarged the possibility of space and freedom of human choices about what information one can generate, process, and share. Information *as* reality created by combinations of bits of information is tantamount to virtuality because it approaches rivalry and replacement of reality.

To facilitate the explanation of how GSE organizations manage information with technological and human resources in order to manage uncertainties and achieve GSE coordination, we turn to information processing theory. In the organizational context, the greatest exposition of the relationship between an organization and information began with Jay R. Galbraith's theory of organizational information processing (OIP).[5] The theory assumes that an organization must design a physical and logical structure that is capable of processing information in order to manage uncertainties it faces. It also assumes that different tasks create different information management needs for the organization. Faced with uncertainties, organizations reduce the need and/or increase their capacity for information generation, processing, and sharing. They reduce the need by pre-planning self-contained tasks and slack resources to continue task execution. They increase their capacity by investing in information systems and by creating lateral relations that enhance flexibility and adaptability using, for instance, teams and task forces. Researchers who are concerned with the relationship between technology, information, and organization have drawn it from OIP theory and its variants such as media richness and information processing.

Media richness theory is Richard Daft and Robert MacIntosh's extension of OIP theory to explain the relationship between technology, information, and organization in terms of variability and analyzability of tasks. Task variability refers to the amount and "frequency of unexpected or new events that occur in the conversion process."[6] Task analyzability refers to the number of exceptional actions in the task and amount of time required by workers to respond to the exceptional events that arise from unexpected events. Unexpected events are sources of uncertainty that refer to the absence of needed information or to a person's inability to predict information about an organizational task. In the face of uncertainty, information management is achieved by generating more information through informal communications, flexible ad hoc communications, and horizontal coordination by mutual adjustment. For instance, because software engineering is replete with high degrees of uncertainty, it requires more information management. Exceptional actions are required as information management behaviors to manage information that is available but equivocal. Equivocal information refers to ambiguous information borne of various and inconsistent interpretations of an organizational situation. Therefore, task variability and analyzability are tightly related with uncertainties and equivocalities respectively (Figure 5.1).

At one extreme end are tasks that are lowly variable and highly analyzable which are routine or simple because they require information management without exceptional actions (e.g. assembly line work). At the opposite end are highly variable and lowly analyzable tasks, described as non-routine because they require information management through exceptional actions. GSE is non-routine work

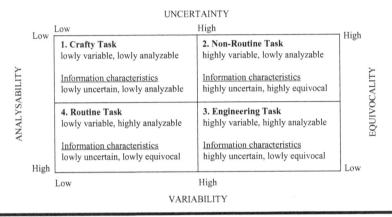

Figure 5.1 Task types and organizational information management.

design or configuration because of the high emphases on teamwork, high reliance on engineers' intellect, fewer routines, and high degrees of coordination by feedback and mutual adjustments. These emphases are typical instances of organizational design of lateral relations to increase capacity for information management. Investments in ICT by GSE organizations for the creation of global virtual teams exemplify increasing capacity for information management. Besides, emphasis on software and task modularization based on Conway's law is a typical instance of organizational design to reduce the need for information management across units or subunits.

Organizational research literature is replete with constructs of coordination that allude to uncertainties and how they can be managed. But uncertainties in research and development (R&D) tasks such as GSE are greater because of their indeterminate processes and outcomes which are further due to task complexity. For this reason, Michael Tushman and David Nadler also draw from OIP to argue that the means for reducing uncertainties in order to achieve coordination in R&D tasks is organic and mechanic information management.[7] Their proposition is that "as work-related uncertainty increases, so does the need for increased amounts of information, and thus the need for increased information management capacity."[8] Based on this, they explain the diverse uncertainties that require information management in terms of three main sources: the subunit's task characteristics, its task environment, and interdependencies with other subunits.

Their emphasis on task- or work-related sources of internal and external uncertainties suggests that task variety and analyzability are affected by environmental factors. And those factors are determinants of the nature of information management required for coordination. Thus, low analyzability and high variety tasks would engender greater uncertainties, and hence require more organic information management. Contrarily, highly analyzable and lowly variable tasks would engender fewer uncertainties and require more mechanical information management.

GSE work is characterized by low analyzability and high variety – it is non-routine work.[9] This is witnessed by a high emphasis upon teamwork (as compared with individual effort), high reliance on engineers' intellect, fewer routines, and high degrees of coordination by feedback and by mutual adjustment.

The task environment refers to those external factors which are attended to by organizational members. The task environment of a GSE unit would therefore refer to the customers whose requirements have to be met. It is also a source of task uncertainty because it is an area that lies outside the software team's domain of control. Customers' requirements and feedback on prototypes as well as requirements for integration of an intermediate product into a bigger software application are typical generators of task-environmental uncertainties. These generators present serious challenges to collaborative software engineering efforts. Global software projects are replete with these environmental challenges which have induced the adoption of new situational and formal methods to address them.

On the one hand, situational methods such as agile development, including its various strands like extreme programming (a process that delivers software as and when a customer needs it) and sprint (a two- to seven-day get-together of software engineers for a project to focus on it) are typically used to address continuously changing customer requirements. On the other hand, lightweight formal methods such as use case driven development, and the use of Alloy and "Z" notations are also typically used for similar purposes. Alloy notation is a small modeling syntax compatible with graphical object models, easy to read and write, useful for describing structural properties, and amenable to full automatic semantic analysis.[10] Essentially, however, the more dynamic the task environment, the greater and more diverse the uncertainties faced by a software engineering team.

Inter-unit task dependencies are another source of uncertainties because an organizational unit's software engineering outcome normally has to be integrated into a larger application. One of the main causes of these uncertainties is the difficulty experienced by geographically dispersed interdependent units as they try to initiate flexible ad hoc communications. When the focal software engineering subunit depends on other subunits to get work done, the greater the degree of instability brought by the other subunits, the greater and more diverse are the degree of task uncertainties.

Task characteristics, task environment, and inter-unit task dependencies are all predominant sources of uncertainties in collocated software engineering. However, the distance factor in GSE alludes to a fourth source of uncertainties: cross-site dependencies. This additional source substantiates the premise that greater and more varied uncertainties characterize global software engineering projects.

Michael Tushman and David Nadler's version of information management theory suggests that organic and mechanical processing of information for managing R&D tasks are closely interrelated. They ascribe the closeness to the fact that repetitive and recurring organic processes can easily be transformed into mechanical versions while the breakdown of the mechanical processes leads to the reformulation of new organic processes or amendment of old ones. They also suggest that the two processes should not be perceived as discrete variables which form a dualism. Rather, the transformations and reformulations suggest the possibility for activities that represent aspects of both processes. Other R&D scholars argue that because software engineering is replete with high degrees of uncertainties, it requires more organic information management. To their well-known examples such as informal, flexible, and ad hoc verbal communications, GSE scholars add identity construction for cross-cultural management via cultural sensitivity, identity construction, media choice, knowledge transfer, building critical social ties, trust, and traveling for face-to-face meetings as key organic processes required for reducing uncertainties. However, the continuous globalization of software projects in recent years and their increasing dependence on interaction technologies seem to be defying this need. This is because software organizations, in the spirit of process virtualization, are substituting key aspects of organic information management for mechanical versions such as ICTs.

This substitution points to the plausibility that, in coordinating global software projects, organic processes may not be too necessary for reducing uncertainties. However, it is also plausible that organic processes are very necessary and are indeed playing an unexplained yet predominant role in managing information to address problems of uncertainties and interdependencies. But these plausibilities, which are embodied in an understanding of the relationship between information management and coordination, remain under-researched in GSE research. This is in spite of the general realization that the success of global software projects depends so much on the pooling of both human and ICT resources to coordinate activities.

Information Management and GSE

How is the problem of emergent, increased, and varied uncertainties in GSE coordination addressed through information management? The analysis undertaken here evaluates the information management capabilities of ICT resources as well as human resources and activities used for GSE coordination. The aim is to demonstrate how technology and human resources contribute to managing the four different sources of emergent, increased, and varied uncertainties in this work configuration for the achievement of coordination. Table 5.1 shows the relationship between uncertainties on the one hand, and technology and human resources for information management on the other (deployed for GSE coordination).

Technology resources refer to the repertoire of technology media installed and used for coordination. These technology media have diverse communication and information characteristics. Some technologies such as group support systems (GSS) support synchronous communications but may not store information. E-mail systems and software bug management systems (BMS) are unobtrusive to engineers' activities but can store information. Instant messaging systems (IMS) are

Table 5.1 Relationship between Uncertainties and Information Management

Information Management through Human and Technology Agency			Types of Uncertainties in GSE	
Factor		Task Characteristics	Task Environment and Inter-Unit Dependencies	Intra-Unit (Cross-Site) Dependencies
Factor	**Quality/Diversity**	Task variety Non-immediacy of query response Unavailability of interlocutor Untraceable communications Spontaneous communications Formality of communication	Changing requirements Eleventh-hour requirements	Varying communication preferences Mutuality of awareness and knowledge Mutuality of understanding
People	**Agility and Experience**	Address task variety	Experience increases expectation; agility facilitates resolution of changing requirements Experience increases expectation; agility facilitates resolution of eleventh-hour requirements	
	Continuity, Longevity, and Relationship Development	Increase informal interactions	Enhance collective agility Learning enhances more efficient and effective ways of resolving eleventh-hour requirements	Awareness of others' communication preferences Mutual understanding through relationship development
	Mobility	Face-to-face encounters enhance informal communications, socialization and relationship development engender spontaneous communications avail interlocutors to each other		Formal face-to-face meetings facilitate mutual awareness Informal face-to-face meetings enhance mutual understanding
Communication Mode (Incorporating Information and Technology)	**GSS** Synchronous Ephemeral Broadcast Unobtrusive	Clarifies task variety Facilitates immediate response to queries Induces informal communications Induces informal interactions	Facilitates notification and collective discussions to resolve changing requirements Facilitates task allocations to resolve eleventh-hour requirements	Also has instant messaging, and document sharing and editing facility that facilitates various communication modes Facilitates task verifications Brings all engineers to the "same page" more efficiently

(Continued)

Table 5.1 (Continued) Relationship between Uncertainties and Information Management

Information Management through Human and Technology Agency	Types of Uncertainties in GSE		
	Task Characteristics	Task Environment and Inter-Unit Dependencies	Intra-Unit (Cross-Site) Dependencies
E-mail Asynchronous Persistent One-to-one and broadcast	Clarifies task variety Helps resolve problems that do not require immediate responses Addresses non-availability of interlocutor Facilitates traceable communications Facilitates formal communications	Supports broadcast of teleconferences scheduled to resolve changing requirements Facilitates task allocations for resolving eleventh-hour requirements in the absence of teleconferencing	Facilitates mutual awareness at both personal and collective levels Brings all engineers to the "same page" less effectively
BMS Asynchronous Persistent Broadcast Unobtrusive	Task variety Broadcasts new bugs, priorities, severities, and assignments Addresses non-availability of interlocutor Facilitates traceability Facilitates formal interactions		Facilitates mutual awareness of bug fixing, priorities, severities, and assignments Formalizes bug-related information through categorizations; facilitates sorting by categories
Instant Messenger Synchronous Ephemeral One-to-one and broadcast Obtrusive	Facilitates immediate response Notifies availability Potentially facilitates traceable communication Facilitates spontaneous communications		Facilitates personal-level mutual awareness Facilitates personal-level mutual understanding
Telephone Synchronous Ephemeral One-to-one Obtrusive	Facilitates immediate responses to queries Facilitates spontaneous communications		Facilitates personal-level mutual awareness Facilitates personal-level mutual understanding

obtrusive and can support synchronous communications. Johann Ljungberg and Carsten Sørensen have organized these diverse communication and information characteristics of technology media into an information exchange framework.[11] The main characteristics in their framework are synchronicity, mode, obtrusiveness, and life. These characteristics are used to analyze the predominant technology media used for GSE work.

Synchronicity is about whether or not communication is concurrent in terms of sending and receiving information. Obtrusiveness is about whether or not communication prompts, aurally and/or visually, the interlocutor about the arrival of information. Mode is about whether communication is one-to-one or broadcast. Life is about whether information communicated is persistent or ephemeral. Analysis of technology media used for GSE work with each of these parameters shows the diversity and richness of technology media and characteristics in this work configuration; thereby underscoring the virtuality approach to the study of coordination (see Table 5.2). The diversity and richness of technology media and characteristics are useful because they are suitable for managing emergent, increased, and varied uncertainties. Only variety can regulate variety according to the law of requisite variety: "the variety within a system must be at least as great as the environmental variety against which it is attempting to regulate itself."[12] The wealth in diverse GSE technology media and characteristics is in the flexibility they afford software engineers for generating, processing, and sharing information at appropriate places and times.

The affordance of flexibility manifests to enable information management because the media are essentially electrical. Recall that electrical media are characterized by high speed, and, because of their materiality, they generate space and time synchronicity between action and reaction. They also generate total field awareness by integrating people, task, and information in a dialectic manner. Moreover, digital information, which is also electronic, is the dominant input and output of software engineering. Thus, we have reason to question Michael Tushman and David Nadler's description of the use of ICT media for information management as "mechanistic information management."[13] ICT infrastructures are indeed mechanical but that is only in terms of their meager mechanical parts. They are dominated by electricity in essence and function; therefore, when a diverse repertoire of ICT media is deployed for information management, affordance of flexibility is expected. The affordance of flexibility of ICT media naturally complements software engineers' face-to-face communications, use of their competencies and experiences, their collaborative relationship development, and their travels and interactions. These are software engineers' human resources and activities which Tushman and Nadler label as "organic information processing."[14]

Human resources and activities are important requirements for information management in the face of emergent and increased uncertainties. When managers of GSE teams realize that task complexity, distance, and technology can undermine collaboration, they structure their teams to reduce the negative effects of these

Table 5.2 Various Characteristics of Technologies, Communications, and Information

		Obtrusiveness		Life		Synchronicity	
		Obtrusive	Unobtrusive	Persistent	Ephemeral	Asynchronous	Synchronous
Mode	One-to-one	Telephone IMS	E-mail	E-mail	Telephone IM	E-mail	Telephone IM
	Broadcast	IMS	GSS E-mail BMS	E-mail BMS	GSS IMS	E-mail BMS	GSS IM
Obtrusiveness	Unobtrusive	–	–	E-mail BMS	GSS	E-mail BMS	GSS
	Obtrusive	–	–	–	IMS Telephone		Telephone IM
Life	Persistent	–	–	–	–	E-mail BMS	
	Ephemeral	–	–	–	–	–	Telephone IM GSS

parameters on coordination.[15] Because of the importance of information management using human resources and activities to achieve coordination, GSE project managers structure the engineering task in order to enable engineers to generate, process, and share information efficiently and effectively. The explanation for this structuring motive is two-fold. First, emergent and increased task uncertainties due to changes in customer requirements up to the eleventh hour that requires high task analyzability by engineers give GSE teams a complex structure. Second, according to Melvin Conway's law, there is a positive relationship between software product structure and the structure of the team engineering it.[16] Therefore, project managers must arrange engineering teams in such manners that will exploit their dexterity and innovativeness in their engineering and communications. Project managers must also complement the arrangement of engineers with diverse ICT media in order to enhance information management and achieve coordination.

Thus, when a project manager collocates a task component at a site, and another component at another, he reduces cross-site task and information dependencies in the task component. Reduced cross-site dependencies and increased site dependencies increase information management with human resources and activities. This is because collaboration is a function of their communication which is, in turn, a function of the team structure. But reduced cross-site dependencies become another coordination problem because the software components must be integrated. To manage the reduced cross-site dependencies and related uncertainties due to collocation of components and development, project managers provide GSE teams with diverse ICT media which are frequent and flexible. Frequent and flexible interactions are useful for increasing mutual awareness of work status at different sites. The frequency of teleconferences held by GSE teams to bring each member to the "same page" and the use of other technology media for cross-site exchange of text, voice, and images all testify to the importance of frequent and flexible interactions for coordination. This mode of information management is in harmony with Joseph McCann and Diane Ferry's suggestion that greater amounts of information exchanged per unit time, added to high frequency of interactions and varied information representations, reduce uncertainties related to dependencies.[17]

Site and cross-site collaborations by global software engineers are not the only human activities for information management. They also use their competences to manage information effectively. An instance of global software engineers' competence is their ability to work with both formal and agile methods of software engineering in the face of emergent and increased uncertainties. This information management ability and actuality accord with Barry Boehm and Richard Turner's suggestion that agile development must be balanced with the required formal discipline.[18] GSE teams work with formal methods in order to satisfy high level organization requirements for documentation of their engineering processes, but they stick to these requirements only in respect of the bare essentials of documentation. Engineers' competences for balancing agility with formal discipline helps them to achieve high responsiveness to changing customer requirements and organizational formal requirements. Continuously changing and eleventh-hour requirements call for the adoption of agile methods. But concerning this demand, engineers' competences in agile development are limited without the ICT media that afford frequent and flexible communications. For example, a GSS technology that affords both synchronous broadcasting of eleventh-hour requirements and task allocations enhances the agility and problem-solving capacity of GSE teams.

Beyond software engineers' collaborations, agility, and competences, their experiences in software engineering after many years constitute a requisite human resource for information management and coordination. Given the challenges with task complexity, eleventh-hour requirements, distance, and technology that face GSE teams, engineers' past experiences are critical. The research by W. Fong Boh and associates on learning from software development experience is supportive of its value.[19]

Beside these human resources, frequent interactions, relationship development, and mobility are the other important activities conducted by GSE teams to reduce uncertainties. Engineers' continuous relationship development and mutual learning lead to high degrees of mutual understanding over time. Developing mutual understanding among distributed team members is a learning process supported by psychological contracts between them that should result in shared mental models.[20] In learning about others, the same process entails some unlearning because some preconceived notions will be challenged by new knowledge. The learning process that leads to high levels of mutual understanding among engineers is also supported by ICT, but technology-mediated learning is limited in the face of the need for organic information management.

Traveling across sites, especially at the early stages of projects, is therefore very important for obtaining and sustaining high levels of mutual understanding in cross-site interactions. This underscores the idea that remote mobility is an important activity that facilitates both organic and mechanistic information management. Traveling to meet face-to-face with colleagues in another site is crucial for reducing misunderstandings in technology-mediated communications subsequent to such meetings. The importance of face-to-face meetings among engineers, especially in early days of GSE projects, is corroborated by the works of Julia Kotlarsky and Ilan Oshri.[21]

The foregoing descriptions indicate that human agency, which pertains to organic information management, complements technology agency in spite of distance and increased dependence on interaction technologies. However, the descriptions of these agencies do not yet explain how they relate to each other for the achievement of GSE coordination. To address this problem, we first relate instances of human and technology agencies to the three main sources of GSE uncertainties: task complexity, distance, and technology. These sources focus analytical attention on particular information management mechanisms and processes for resolving tasks, bridging distance, and normalizing technology respectively. These particular information management processes and mechanisms are employed for this analysis because they are the main instruments used in GSE across the world. This analysis is important because information management can be consciously employed or be deemed as chance outcomes or be lost depending on researchers and project managers' comprehension of them. The aim here is to enhance our comprehension of it and, hence, to dispel any idea of chance, so that it can be consciously employed for coordination.

We should make a distinction at the outset between human resources and activities that are requirements for software engineering tasks on the one hand, and those that were requirements for information (about tasks) on the other. Based on this distinction, we can note that experience, collaboration, and agility are task resolving requirements because they directly concern efficient responsiveness to task challenges. Experience and collaboration are exercised without and are not enhanced necessarily by interaction technologies. Agility, however, is exercised with and is enhanced by these technologies. Experience has virtually no relationship with distance, while collaboration and agility have negative relationships with it. Experience and collaboration, coupled with the interaction technologies, enhance engineers' agility, which often leads to efficiency in response to eleventh-hour requirements.

Thus, on the one hand, experiences point to the importance of *selection* of engineers to form teams according to their pure task-resolving resources. On the other, their collaboration points to the importance of task *structuring* according to the need to deal with task complexity. Selection and structuring therefore emerge as information management foundations for GSE coordination. These foundations further suggest that being able to coordinate GSE projects may not have to do with merely deploying interaction technologies to connect remotely distributed expertise. It has much to do with adjusting distance by collocating task components. It even has much more to do with acquired human qualities such as experience, and the experience could be in a general application domain, or in a methodology area such as extreme programming depending on the circumstance. As more technology is deployed to support GSE, more human resources and activities which are not technology-dependent are required for coordination.

The distinction between tasks and information (about tasks) also points to frequent interactions, continuity/longevity, and mobility as requirements for reducing distance-related information uncertainties. This is because they concern information exchanges leading respectively to mutual awareness, learning, and understanding, which are important outcomes but do not directly address task challenges. These requirements (frequent interactions, continuity/longevity, and mobility) have distance-bridging potential because they can reduce the negative effects of distance on mutual awareness, learning, and understanding, thereby making them more interesting for prognosis. Furthermore, among these requirements, frequent interactions are directly technology-dependent while continuity/longevity and mobility are not. Yet continuity/longevity and mobility complement frequent interactions significantly, leading to effective information management that overcomes the negative effects of distance on mutual awareness, learning, and understanding. This understanding of continuity and mobility, which reflects human agency because they are not technology-related, also points to the importance of *exploitation* (not in a derogatory sense) of engineers' distance-bridging activities as an additional information management requirement for GSE coordination. This implies that exploitation is more proactive than just leaving remote interlocutors to build relationships through their technology-mediated interactions. It is a matter of instituting measures for human agency to flourish and supplement technology agency. To wit, as more technology is deployed to support interactions, more human resources and activities (which are not technology-dependent) are required for mutual awareness, learning, and understanding (Figure 5.2).

Carl May and Tracy Finch's normalization process theory helps to deepen our understanding, evaluating, and thinking through the processes of implementation, embedding, and integrating technology into work processes.[22] The way to understand the normalization of technology in GSE projects is to comprehend the relationship between its objective and subjective properties. On the one hand, its objective properties can be understood in terms of variety in synchronous/asynchronous, obtrusive/unobtrusive, and broadcast/one-to-one communication options, as well as persistent/ephemeral information life options. On the other, instances of its subjective properties can be understood in terms of how the objective properties, together with others such as ease of use, shape the perceptions of its users. Thus, users perceive technology as normalized in their work if it has so disappeared from their consciousness that their reactions to it reflect sub-consciousness or if it is no more drawing considerable attention as it used to do.

This normalization occurs in GSE teams because of the diversity in the installed technologies, of the flexibility in their uses, and of how the engineers take advantage of those features. They take advantage by using them to support their agility,

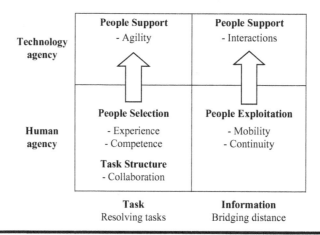

Figure 5.2 Information management matrix.

mutual awareness, and learning. But the process of normalization begins with the selection of human resources and the exploitation of their activities. These are crucial foundations that enhance the technology support for agile development and learning in frequent interactions. Thus, by the time high levels of these organic processes are achieved by a team, their technology would be normalized. The implication of technology normalization as a teleology of technology-mediated communications is that it points to the importance of *supporting* organic processes with technologies to manage uncertainties and interdependencies. Technology normalization also subsumes the importance of selection and exploitation of human resources for enhancing the coordination process.

By emphasizing uncertainties and information management, this chapter goes beyond existing perspectives such as knowledge, communication, process steps, and architecture plans. It explains how the range of human and technological resources can be mobilized and applied to address the problem of uncertainties. It also demonstrates how the functional relationships between the resources address the problem and achieve coordination. Based on these explanations and demonstrations, this chapter contributes two main information management ideas to GSE coordination theory.

First, it contributes an understanding of coordination that is more holistic than previous ones. As far as support for interactions in GSE is concerned, this perspective resonates with notable ones such as shared knowledge among team members, time-separation effects, task- and product-related methods, and software architectures, plans, processes, and ad-hoc communications. At the same time, this perspective is different because it conceptualizes selection and exploitation as important foundations for organic information management. These foundations have been taken for granted in existing perspectives whereby the exegesis of coordination is confined to support for interactions, while the implications of their antecedents – selection and exploitation – are overlooked, rendering them substantially limited. The following are notable instances of existing models and their limitations.

According to J. Alberto Espinosa and his colleagues, there exists a shared team knowledge perspective on coordination which is defined by constructs such as shared knowledge of task and team, and task and presence awareness.[23] The importance of shared knowledge, task awareness, and presence awareness resonate with the need for reducing uncertainties. The shared team knowledge perspective is therefore useful for understanding coordination, but it does not explain well how its fundamental constructs can be generated. Its explanations of shared knowledge center largely on support for interactions with very little reference to how the increased and diverse uncertainties can be addressed by shared team knowledge. It is, therefore, limited in its effectiveness for analyzing how GSE work is coordinated.

The time-separation perspective on coordination, developed by J. Alberto Espinosa and Erran Carmel,[24] is also captured in this information management perspective under how intra-unit interdependencies are negatively affected by time differences. In the time-separation perspective, time delays in communications across geographically distributed sites increase coordination costs. This point is useful because it shows how the total cost of carrying out a task is influenced by the cost and effectiveness of different communication mechanisms and by time-separation delays. However, it is also limited only to the effects of time-separation, which are esteemed as factors that compound the negative effects of geographical separation. The time-separation model precludes other factors such as task complexity and interdependencies which are additional sources of uncertainties and can also compound these negative effects. This perspective rightly relates to the kinds of support for communications that can reduce the coordination costs emanating from time-separation. But it undervalues how the costs of time-delays can be reduced through the selection of human resources and exploitation of their non-technology related actions.

Rebecca Grinter and her colleagues' geographical perspective on coordination which deals with distance in R&D work also resonates with this information management perspective.[25] Constructs such as expertise and product structure reflect respectively competence and product components in the information management perspective. Both perspectives are in agreement that these resources should be co-located to facilitate ad hoc communications and collaboration because they both espouse support for communications by structural re-arrangements and by depending on engineers' expertise. However, their perspective addresses the problem of distance, and not necessarily other problems such as task complexity and increased deployment of information technologies which are equally significant sources of uncertainties. As a result, they overlook the exploitation of human activities such as mobility and continuity as profitable for dealing with distance in R&D work. All these limitations of their perspective have been addressed in this chapter.

Moreover, the proposal by James Herbsleb and Rebecca Grinter that ad hoc communications are complementary processes that can address the challenges posed by distance is in harmony with the argument that organic information management is a crucial form of support for interactions among GSE team members.[26] However, their proposal does not detail any elements of ad hoc communications and their interrelations. It is also not based on a comprehensive study of the increased and diverse uncertainties, and so it overlooks the profitability of selecting human resources and exploiting their activities. As a result, it is limited in its capacity to help researchers and managers to understand organic information management requirements for coordination.

Second, this chapter argues that the information management perspective on coordination is more appropriate for studying the combination of GSE work, interactions, and ambiguous change therein. Existing perspectives all show that researchers have assumed quite a static relationship between the problem of increased uncertainties and coordination. Admittedly, they have rightly viewed GSE coordination as a process involving communications and collaborations. Interestingly, this is a simple-process view. The information management perspective shows that the relationship between uncertainties and coordination is more dynamic. The dynamic relationship surfaces in changes in task characteristics, task environment, technology, and related dependencies.

Thus, the information management perspective shows that the nature of uncertainties in GSE necessitates the perception of coordination as a meta-process involving changing task, selection, interaction, exploitative, and learning processes within the existing knowledge, communication, and collaboration processes. In terms of research, it advocates a grounded theory methodology that will enable a more insightful study of all the changing circumstances that are sources of the problem of increased uncertainties. This perspective is a better study guide because it can enable researchers to identify known and emergent sources of uncertainties as well as their variations. In particular, it foments the analysis of phenomena such as task, technology, distance, and changing requirements, as well as the structuring, selective, exploitative, and supportive actions enacted by project managers and software engineers, all in one single effort. Practically, it is also a better guide for managers in terms of how information management interventions such as hiring, acting, interacting, and learning in changing circumstances can be applied responsively for coordination to be achieved.

In short, existing information management perspectives of GSE coordination and the appraised one in this chapter are in agreement that organic information management is an important requirement. But while existing ones end their explanation at that level, this chapter proceeds beyond it to detail the particular constituents of organic information management requirements and to model their interrelations. Existing perspectives are not sensitive to the new sources of uncertainties that challenge traditional information management requirements for coordination. But this model has been sensitive to these increased and diverse uncertainties because it conceptualizes coordination not merely in terms of supporting software engineers' interactions but also in terms of selecting their task-based resources and exploiting their distance-bridging activities.

Chapter 6

Exploitation of Geography

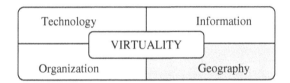

These days, many software organizations are outsourcing some of their operations to countries such as the Republic of Ireland, India, China, and some English-speaking African countries. One predominant aim is to take advantage of cheaper skilled labor which has become expensive in many "home" countries; another aim is to practice 24-hour software engineering. This trend depends largely on technological innovations as well as information networks and flows that are espoused by Manuel Castells in his theory of the network society.[1] Jannis Kallinikos also sees the GSE trend as "the quest for alternative economic and organizational practices combine[d] with the impressive instrumental involvement of information and communication technologies."[2] Castells prognosticated that such technologies would overcome distance, time, and structural barriers, and organizations would benefit from 24-hour continuous work progress in different offices around the globe.

The notable dimensions of global software engineering (GSE) and its coordination are geographical, cultural, temporal, political, technical, and historical; but the geographical dimension is foundational because it induces the emergence of the others. The geographical perspective on coordination is critical because geographical distribution of software engineering resources and activities has great potential to undermine effective dependencies among software engineers. The global distribution creates new problems of multi-site work coordination.[3] Furthermore, notable concepts such as "nearshoring" and "offshoring"[4] confirm that geography really matters in both its problematization and conceptualization in coordination theory development.

The GSE literature is replete with the ontology of geography where place and space are acknowledged as predominant aspects of team composition, resource distribution, task structure, and product marketing. The word global itself carries heavy overtones of the ontology of geography as a predominant aspect of this work configuration. The word global describes software engineering, but the description is more than a color. The word global is a description of an integral aspect of the problematization of coordination by software engineering researchers. For instance, global spatial distance between software engineers is universally acknowledged by these researchers, and almost all publications on GSE coordination include spatial distance in problem discussions. Problems such as mutual misunderstanding, cultural incompatibilities, task conflicts, interpersonal conflicts, and uncertainties are all discussed with strong references to geography.[5]

Unfortunately, geography has been confined to problematization at the expense of conceptualization; and so software engineering researchers currently have a weak theoretical understanding of the geography of GSE coordination. Geography is ontologically dominant in extant coordination publications but epistemologically negligible. Wherever coordination theory has been developed in these publications, geography is presumed rather than explained explicitly. The geography of coordination is about the *where* of the distributed resources – people, tasks, information, and technologies; but it is yet underexplored and not well-understood. What have been espoused all the while in the literature is the *what* and the *how* of these resources. Without a clear conceptualization of the geography of coordination, researchers and managers will have to rely on the existing what and how ideas which, at best, will only provide us with intuitive guidelines.

Now that the GSE work configuration has become a predominant mode of software engineering, a conceptualization of space and place dimensions of coordination in relation to existing ones is important to enhance our understanding of coordination. Moreover, since places and spaces are resources, there is a need for an explanation of how geography is exploited to achieve coordination in order to brighten its epistemological light. This book takes up this challenge. In Chapter 4, there was a hint of the coordination value of geography where the combination of global spatial distance and group support systems was discussed as a strategic opportunity for information management leading to coordination. With reference to the logic of virtuality framing this study, information and communication technology (ICT) media enlarge the possibility of space and freedom of human choices of what information one can generate, process, and share. ICT media complement spaces and places by the totalizing effect of field awareness through information speed, synchronicity, and richness – what McLuhan calls the "cosmic consciousness" of the human mind. Cosmic consciousness among global software engineers is the continuous experience of closeness and interdependence in spite of and through spaces and places, thereby underscoring the global village metaphor.

Given this background, knowledge about how geography is exploited to achieve coordination is important both theoretically and practically. Theoretically, this chapter explains how and why space, place, and ICT combine to address problems of task dependencies. Based on the explanation, researchers can analyze salient GSE coordination problems such as integration efforts, ethnocentrisms and conflicts, communication and technology choices, and developers' behavioral dynamics. Practically, the explanations provide guidelines for managers to draw upon in order to address the persistent coordination problems of task structuring, global expert sourcing, exploitation of developers' distance-bridging qualities, and their interaction support.

Geography and ICT

Place, space, and ICT are parameters that can assume the roles of coordination mechanisms in virtual teamwork. For example, uncertainties in a highly inter-dependent software engineering task may necessitate the need for computational mechanisms across distributed sites to facilitate interactions and knowledge shar-ing. However, interactions and knowledge sharing may be undermined by socio-cultural belief systems that translate into context-bound meanings of information and nature of knowledge called the "mutual knowledge problem."[6] Existing pro-cesses for GSE coordination in the face of these knowledge problems are cross-cul-tural management by negotiating cultural sensitivities,[7] enabling media choices,[8] inducing knowledge transfer,[9] building critical social ties,[10] developing mutual and interpersonal trust,[11] grouping software components and developers,[12] and travel-ing for face-to-face meetings.[13] These processes are undergirded by place, space, and computational mechanisms, but only passing references are made to them by these publications.

The question about the "where" of organizational coordination is understood in terms of place and space which are fundamental concepts of geography.[14] The rela-tionships between these concepts are understood in terms of historical and philo-sophical renditions. There is the pre-modern rendition of the relationship, dating back to the seventeenth century, which proposes a tight coupling between space and time. The understanding of place in this rendition is in the emphasis on its location, boundedness, and strong ties with cyclical time. Corresponding to this understanding of place is the Newtonian view of space as an absolute and real entity. Newton's argument is that the place of a body is the space which it occupies and that the only feasible analysis of true motion of a body requires reference to absolute places, and hence the existence of absolute space.[15] Based on this view, the geographer John Agnew describes space as absolute, concrete, and real.[16] Therefore, tight space-time coupling implies that the pre-modern understanding of space is its metaphysical affinity to place – both are absolutely locational. The organization of human activities underpinned by this view gives priority to place.

Organizations of the Industrial Revolution, driven mainly by mechanical media that process things relatively slowly (as compared with electrical media), exhibited prioritization of place over space. In these organizations, coordination and control were exercised and achieved by managers through placement of resources. One of the main determinants of limitations to coordination and control was the concrete location of resources and activities. Even where organization of human activities was done by globally distributed work, the priority of place over space was the rule. An example is the case of the Hudson Bay Company (1670–1826) reported by Michael O'Leary and colleagues.[17] The Hudson Bay Company (HBC) was a globally distributed organization headquartered in London. It had fur trading posts and 99% of its employees in Canada (which then was a long sea voyage as compared to a short air trip today). HBC exhibited priority of place in coordination and control by changing trading posts and recruiting locals for the trading posts. Place determined coordination and control in HBC because it was the organization and number of posts and employees which were changed from year to year in response to concrete trade challenges.

From a distance, coordination and control from the London headquarters were severely limited by the "weather, hunting and trapping seasons, and the annual passage of the Company's ships, whose access was limited to a brief window in late August when the narrow opening of the Bay was not frozen closed."[18] In the absence of electrical media, the London headquarters had a problem of spatial distance in coordinating and controlling workers in Canada. Therefore, it had to add organization by trust to the place-based coordination and control through its posts: "We know it is impossible at this distance to give such orders as shall answer every occurrence and be strictly observed in all points, so that when we have said all, we must leave much to your prudent conduct, having always in your eye the true interest and advantage of the Company, who have chosen and trusted you in the chief command they have to bestow."[19] Hence, space was regarded in the pre-modern rendition – in the Newtonian view – as an absolute container within which matter is located or a grid within which substantive items are placed.[20]

The alternative Liebnizian view of the relationship between place and space describes space as relative to events and objects. This view is typical of the modern rendition of the relationship between space and place that subordinates place to space;[21] especially because of the influence of electrical media.[22] This rendition is traceable to Albert Einstein's general theory of relativity that explains space and time as expandable and bridgeable. Carol Saunders and colleagues,[23] for instance, draw heavily upon Einstein's relativism theory to develop their own theory of virtual space and place. One of the main premises of their theory, also drawn from Helen Couclelis and Nathan Gale's work on *Space and Spaces*,[24] is their assumption that perceptual and cognitive spaces exist but physical spaces do not exist in virtual worlds. Other scholars who give epistemological priority to space ahead of place, especially those in information systems and computer science, also assume that space is perceptual and cognitive, and therefore unbounded. The perceptual and cognitive descriptors signify the almost infinite opportunities for the mobility of people, information, tasks, and events. Working with this assumption for many years after the Second World War has become an indicator of modernity.

Thus, the major point of departure from the pre-modern rendition (which coupled place and time) was the decoupling of space and time.[25] By this decoupling, space as a relative concept assumes a new status of a limitless phenomenon that gives freedom for the movement of activities and resources across the world, while place retains its status as a bounded phenomenon that limits them.[26] This rendition prioritizes space ahead of place, but maintains that both of them are relative concepts. Place is viewed as an anachronism which must give way for space-spanning connections and flows of information, things, and people to signify modernity. Accordingly, in the early years of the advent of the world wide web and the popularization of the internet, some researchers, notably Frances Caincross,[27] sounded the death knell of distance because of the power of ICT to overcome spatial barriers. This reasoning has been a powerful influence on the virtualization of teamwork. Both the theory and practice of teamwork virtualization envision the digital and indexical representation of workers and their work to enable remote interactions across space,[28] and to enable knowledge transfer through transactive memory.[29] Related to this reasoning, many strategic information systems models also assume that the use of ICT should enable organizational activities that overcome both spatial and temporal distances. Lately, GSE and other forms of virtual teamwork have emerged as phenomena built upon the same assumption that the combination of space and ICT creates a limitless virtual space to workers. But the context of virtual worlds made of digits is ontologically different from virtual organizations which include a lot of physical things including places.

The economic interpretation of the modern rendition is that place was a local, traditional, static, and nostalgic phenomenon that gave meaning to the feudal and agricultural economics of the past.[30] This conception is contrasted with the global, modern, mobile, and radical understanding of space that has given rise to new patterns of capital accumulation and wealth generation. The modern understanding of space underpins the global socio-economic transformations witnessed in the Enlightenment, the industrial revolution, and the information revolution. These transformations are traceable to scientific and technological advancements that have manifested in electrical media. The heavy influence of electrical media is evidenced by instances such as scientific management, mass production and consumption, virtual teamwork, electronic commerce, ubiquitous and cloud computing, and business analytics with big data. All these instances exhibit the totalizing effect of field awareness through speed, synchronicity, and richness which stimulates the mind – that is, cosmic consciousness. But the modern rendition underplayed the ontological status of place by over-emphasizing the unbounded character of space.

Currently, the pre-modern and modern perspectives have been superseded by the post-modern version that has brought place back into geography and economics. The pre-modern perspective of place emphasized what John Agnew calls its geometric conception which refers to "nodes in space simply reflective of the spatial imprint of universal physical, social or economic processes."[31] However, the postmodern version emphasizes its phenomenological conception that is understood in terms of "milieux that exercise a mediating role on physical, social and economic processes and thus affect how such processes operate."[32] In GSE, Marisa D'Mello and Sundeep Sahay, for instance, suggest that the "social meanings and existential significance of humans are related to places (physical, social and electronic)."[33] This phenomenological conception of place presents both relational and experiential understandings of it without sacrificing its physical ontology.

Closely related to this phenomenological conception is Maryann Feldman's geography of innovation theory which assumes that places are as powerful explanations for resource organization as firms are.[34] She speaks of location in terms of "a geographical unit over which interaction and communication is facilitated, search intensity is increased, and task coordination is enhanced."[35] Based on this assumption, the actions of entrepreneurs define the character of places and contribute significantly to economic clustering of innovation in regions. Location-based interaction and communication translate information into knowledge required for innovation through a non-linear process. A typical example is the number of ICT companies that have formed an innovation cluster in Silicon Valley in the United States. The prevalence of the innovation cluster there has inspired the formation of technology parks in many countries around the world. Thus, even for high technology innovation organizations which are dominated by electrical media, the phenomenology of place is a crucial factor for their task coordination (internally and externally). Places which are designed and experienced successfully by entrepreneurs as clusters of innovation are characterized by "a spirit of authenticity, engagement, and common purpose."[36]

Places are powerful explanations for resource organization and coordination because they correspond to mechanical media which are not obliterated by electrical media when virtualization of organization occurs. The hearts or headquarters of the most virtual organizations are in buildings which have strong placial identities in addition to their spatial ones. Therefore, virtual organizations have combinations of mechanical and electrical media in various proportions. For this reason, virtual organizations still practice Alfred Marshall's agglomeration economies which are informed by mechanical media principles. It is true that agglomeration economies pertain primarily to clustering of organizations in particular places. But the principle of external economies such as exchange of ideas and conversion of information into knowledge applies to both localized firms and localized individuals. For localized individuals at particular sites in GSE economize software production with mechanical media such as paper, buildings, photocopiers, telephones, and printed words.

Although these mechanical media may now be dominated by electrical (and digital) media, the former are enduring because, as Marshall McLuhan's media ecology indicates, new media combine with rather than replace older media.[37]

The benefit of close-to-market production of software which motivates the formation of GSE teams is also place-centered. The materiality of place is the central argument in James D. Herbsleb and Rebecca E. Grinter's study of how geographic separation of software engineers and their tasks undermines informal communications that are required for coordination. From their study, they made the following observation: "Architectures, plans, and processes are all vital coordination mechanisms in software projects. However, their effectiveness extends only as far as our ability to see into the future. Handling the unanticipated both rapidly and gracefully requires flexible ad hoc communication. This need became clear as we examined how distance interfered in a variety of ways with the project teams' effective communication."[38]

Therefore, in seeking to understand how geography is exploited by GSE organizations for coordination, this book aligns with the phenomenological conception of place which presents its relational and experiential understandings of it. The phenomenology of place in the post-modern rendition complements the generality of space espoused in the modern rendition. Thus, places are parts of spaces, place implies space, and space is a property of the natural world but it can be experienced in particular places.[39] The combination of space and ICT provides a virtual concept as well as practical opportunities for free movement of digital data; place is a phenomenological yet real locality invested with social meanings of appropriate behavior and cultural expectations, and related to social and economic needs (Figure 6.1).

Researchers, especially those in ubiquitous computing and human–computer interaction sub-fields of research, adhere to the post-modern perspective because they believe that places and spatial distances between them indeed matter significantly. Their designs are aimed at computing that relates to places. For instance, at the heart of ubiquitous computing lies context-aware computing which assumes that computing can leverage aspects of particular places and provide valuable digital information services to the user.[40] User-centered design of ICT suggests that the user, the technology, and particular place-based information together define usability. This is witnessed in computing with global position satellite (GPS) systems and "smart" domestic appliances.[41] The same reasoning applies to prospects for collaborative virtual environments evidenced by computer-supported collaborations between multiple users of ICT.

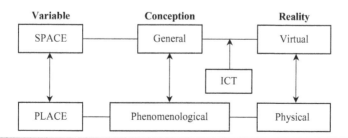

Figure 6.1 Complementary relationships between space, place, and ICT.

Geography and GSE Coordination

The phenomenological conception of place and the general conception of space (combined with ICT) underpinning the exploitation of geography for coordination implies the analysis of geography, technology, and information together. Our interest in exploitation of geography means we are concerned with the physical, relational, and experiential aspects of places, spaces, software engineers, information, and technology. Recall (from Chapter 5) that GSE is a non-routine or complex work configuration because of the high emphases on teamwork, high reliance on developers' intellect, fewer routines, and high degrees of coordination by feedback and by mutual adjustments. These emphases are typical instances of organizational design of lateral relations to increase capacity for information processing. Investments in ICT by GSE organizations for the creation of global virtual teams exemplify increasing capacity for information processing. Besides, emphasis on software and task modularization based on Conway's law is a typical instance of organizational design to reduce the need for information processing across units or subunits. As James Herbsleb and Rebecca Grinter point out, distance is as important as architectures, plans, technologies, and information for the achievement of GSE coordination.[42] Therefore, the analysis of the exploitation of geography is informed by assumptions from theories of geography of innovation (GoI) and organizational information processing (OIP).

OIP is an appropriate complementary analytical framework to GoI for explaining the role of geography in organizational design because of four additional reasons. First, pre-planned and flexible physical designs for information management correlate directly and respectively with coordination mechanisms and processes which this research will explore in terms of place and space. Second, software engineering is both an information management and innovative activity. Information management and innovation combine to produce many exceptional events that require exceptional actions including geographical adjustments in resource locations. Third, OIP proffers ideas about how dependencies and uncertainties pertaining to organizational activities and information resources can be placed to achieve coordination. Fourth, since place and space in coordination are being explained from a software engineering perspective, its information processing character corresponds with OIP and enables the assurance of construct validity. The appropriateness of OIP for studying virtual teamwork is further demonstrated in its application to existing research on coordination in information systems development outsourcing,[43] on project managers' practical intelligence in software offshore outsourcing,[44] on operational agility development,[45] on virtual design teams,[46] and on transactive memory in virtual teams.[47]

There are two main interrelated design decisions made by GSE organizations to process information that achieves coordination. First, software components and developers are collocated at individual sites to enhance developers' focus and performance. Second, collocated teams collectively are made to generate, handle, and share information about their task across sites to meet the information management challenges posed by the collocation. The first decision has stronger overtones of place than space because it predicates software engineering coordination on the exploitation of physical location. The second decision has stronger overtones of space than place because it predicates coordination on technological exploitation of space.

Collocating software components suggests that both the high variability and low analyzability of the engineering task are strongly related to the geographical dispersion of its components. After all, the components are collocated in places as an organizational design for information processing that manages variability and analyzability of software engineering. Reducing task dependencies among developers at a site essentially signifies the consideration of place in organizational design to reduce uncertainties and equivocalities. Therefore, when organizations collocate task components and developers, they redesign the organization for information processing. In so doing, they acknowledge that task performance is not only a function of variability and analyzability but also of the phenomenology of place. John Agnew's conceptualization of the phenomenology of place points to how it mediates the "physical, social and economic processes and thus affect[ed] how such processes operate."[48]

In spite of the mediatory role of ICT in virtual teamwork on software engineering that is guaranteed by electrical media richness and synchronicity, full electrical media richness is impossible. This is because of the enduring mediatory role of place in information processing among groups of software developers. "Because we are not physically together," says the Technical Lead of a GSE team spread across the United States and the Republic of Ireland, "I had to design the [components] so that they are completely independent between the regions" (Chapter 9). This implies that the physical separation of the developers and task components has negative effects on exceptional actions such as ad hoc communications required for working optimally on those components. It also implies that if the components are physically scattered across the sites (places), then that scattered placement will negatively affect the software engineering processes.

The phenomenology of place has a generative capacity to reduce uncertainties and increase dependencies among developers at a site. Increasing dependencies enhances information processing because their proximity enables them to tackle the lowly analyzable nature of software engineering. Collocation of software components and engineering according to the phenomenology of place is the organization's placement of software components in accordance with the idea of software modularization by David Parnas.[49] In software modularization, software components are grouped into modular logical and physical units to achieve a more efficient and effective workflow.

The phenomenology of place is a necessary antecedent for information processing because it undergirds the physical grouping of software components and developers in order to generate information management behaviors that reduce uncertainties and equivocalities. Logical and physical modularization is a necessary organization (re)design for information processing that enhances task analyzability and coordination. Therefore, place is a generative mechanism for physical organizational design that enables engineers' information management for GSE coordination.

Physical proximity between the engineers at a site, being an organizational information processing design, foments informal and spontaneous interactions between them. These interactions help to enhance the analyzability of application engineering by significantly reducing communication delays and breakdowns between them. Informal and spontaneous interactions in software component engineering are social processes required for engineering of mutual understanding, shared values, and social ties.[50] Audris Mockus and David Weiss made a similar finding alongside their inverse finding that developers at different sites often suffer inhibited communication and coordination.[51]

The form of information exchanged in these interactions is analog, signifying that it is tightly coupled with analog media and carriers. Place and developers are analog mechanisms, with place being the necessary antecedent structure required for generating information processing behaviors that are useful for managing the constantly changing requirements faced by GSE teams. Place enables team members to process information via sharing which is required for managing the highly analyzable software task. Therefore, place is also a generative mechanism for social organizational design that enables engineers' information management for GSE coordination.

Organizational design that leads to high dependencies at particular sites is important to speed up task performance. James Herbsleb and his associates have noted that loss of rich, subtle, and ad hoc verbal interactions in cross-site software engineering makes the work take longer times than single-site work.[52] His reference to time signifies efficiency in the workflow of both the software task and its components. Workflow reflects the economic aspects of David Parnas' concept of software modularization. One of the core merits of GSE is economic efficiency in software production in order to reduce engineering and time-to-market costs.[53] This means that reducing the time to finish the engineering of an application is an information processing design to manage its variability by enhancing the efficiency of interactions and task performance. Organizational design that includes place as an antecedent structure is necessary for efficient information processing needed to manage variable nature of software engineering. Place is a generative mechanism for high dependencies that lead to enhanced economic processes in software component engineering. Therefore, place is also a generative mechanism for economic organizational design that enhances efficiency of engineers' information processing for GSE coordination.

Because coordination is required for efficiency and effectiveness of software engineering, the role of place in GSE coordination can be explained by its value for these. When organizations give priority to sites or locations ahead of technology-mediated collaboration in order to achieve efficiency and effectiveness, they essentially declare a higher valuation of place than ICT. The high valuation of places and resultant collocation of work for software engineering efficiency and effectiveness are interpreted as *workplace formation* – the formation of software component workplaces. This interpretation draws upon the geography of innovation where it is assumed that entrepreneurs' actions define the character of place.[54] Forming refers to the organizational action to develop innovative clusters at the sites, and this is analogous to molding a pot from a lump of clay.

Without innovative clusters, they invariably reflect the physical, instead of the phenomenological, conception of places characterized by engineering inefficiency ("out of place"). Workplace formation is not bypassing particular sites but rather leveraging them in order to ensure that software engineering work is "in place." Thus, the workplace is characterized by a combination of physical, innovation, and economic dimensions as compared with place which only has a physical character. The workplace functions as a coordination mechanism, defined by Kjeld Schmidt and Carla Simone as an artifact with physical ontology and persistence (bug report form, timetable, etc.).[55] This is because it is formed by organizational action to stimulate interactions that achieve GSE coordination and efficiency.

However, workplace formation does not necessarily yield interactions and effective GSE coordination. For engineers may be gathered there and yet ignore each other if it is not invested with social meanings of appropriate behavior and cultural expectations.[56] Therefore, how workplace formation is implicated in coordination for engineering efficiency requires explanation of how the engineers experience their workplaces. The workplace as designed as an instance of the phenomenological conception of place, pointing to its relational and experiential understandings according to Martin Heidegger.[57] Productive relations between software engineers depend on the socially constructed understanding among them that collaborations and ad hoc verbal communications are appropriate behaviors,[58] and that mutual understanding and social ties are cultural expectations in GSE.[59] Because collocated engineers' collaborations, communications, and mutual understanding are meaningful for software engineering, they are interpreted as having social meanings which are held by engineers.

Organizational design that collocates software components to exploit the engineers' collocation signifies identification of the formed workplaces with social meanings – termed *workplace identification*. Identifying is analogous to tagging a thing with characteristics that distinguish it from others. These social meanings are coordinative protocols which, according to Kjeld Schmidt and Carla Simone, are agreed-to conventions of an ensemble stipulating their responsibilities.[60] Identifying the formed workplace with these coordinative protocols is vital for this coordination method, especially because the protocols are objectified in the workplace mechanism. Protocols are objectified by giving them permanence – sometimes by representing them in concrete form.[61] Protocol objectification enables software developers to experience their sites as workplaces for generating high task and information dependencies. But because experience is ephemeral and interpretive, workplace identification with coordinative protocols and protocol objectification are continuous processes that operate with the persistent workplace mechanism to achieve GSE coordination and efficiency.

If organizations do not design high task and information dependencies at a site solely because two developers are there, then the developers' experience is not solely attributable to the dependencies but also to workplace formation and identification with ad hoc and verbal communications protocol (Figure 6.2). The high task and information dependencies needed for coordination and efficiency of software component engineering is hardly achievable without workplace formation and identification. Thus, place is implicated in GSE coordination because of engineering efficiency; it is implicated by its formation and identification as a software component workplace (Figure 6.2).

In spite of the collocation of software tasks and developers to exploit the phenomenology of place, leading to workplace formation and identification, developers generate, share, and manage digital information across sites. Mutual understanding and sound interrelations are necessary in order to be adequate to the threat of low task dependencies and mutual knowledge problems across sites. It is also necessary to be adequate to the rapidly changing customer requirements that demanded both problem-solving and timely information management from the team. In the face of the threat posed by the particular problem of low task dependencies across sites to team cohesion, developers collaborate with digital media. The operational problems at various sites that have to be resolved and the unstable operations caused by continuously changing requirements are physical process problems.

These problems induce organizations to adopt ICT infrastructure that open up virtual workspaces for electronic collaborations and mutual task awareness across sites. The technologies used mediate the spatially distributed workspace according to its promise to enable teammates to work together without face-to-face contact.[62]

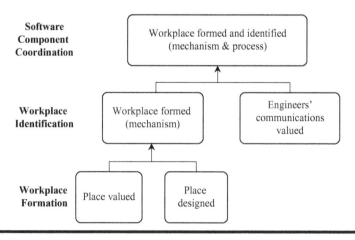

Figure 6.2 How place is implicated in GSE coordination.

This promise underpins the formation of GSE teams because it provides organizations with opportunities beyond location and time. Virtual workspaces enable high digital information dependencies evidenced by continuous digital information generation, handling, and sharing by developers. Such collaborations and awareness constitute means to harmonize task and information about it.[63] Hence, they are means for information processing that organizations design to enhance product analyzability of virtual teams. However, as Paul Dourish suggests, the ICT-mediated world still relies on the physical world within which it is embedded.[64] For this reason, the high dependencies only exemplify the new forms of practice and organizational knowing which he discussed as consequences of the appropriation of existing spaces "in different ways."

In social terms, team cohesion is built through members' continuous electronic collaborations. Thus, organizations deliberately design virtual software engineering teams to create capabilities for information processing that enhance product analyzability in the face of mutual knowledge problems.

In economic terms, developers make efforts in these collaborations to resolve the problems early at a relatively lower cost compared with a later resolution.[65] Furthermore, team members' electronic access to more experienced developers for guidance using virtual workspaces is an economically efficient process that avoids the cost of physical travel. All these economic processes reflect the organizational design of information processing capabilities to enhance product analyzability.

The use of space through ICT is interpreted as utilization of a digital information space (dataspace) that enables free data movement and access for mutual awareness and responsiveness to externally changing requirements – I call this *dataspace utilization*. This interpretation draws upon virtualization of teamwork where it is assumed that workers operate on digital representations of themselves and their work.[66] Virtualization is a representation of the team's space with digital information to enable its workers to interact remotely. To utilize is to work with a part of the vast virtual space generated by ICT. Responding to externally changing requirements results in software engineering effectiveness which is a standard of how well an organization meets external demands.[67]

GSE teams work with the global standards of connectivity in ICT and their diverse communication technologies to utilize their dataspace and achieve engineering effectiveness and creativity. The dataspace utilized is general even though the underlying infrastructure of ICT is particular. It is general because it operates mostly with the global connectivity standards than the particular infrastructure to overcome space and time boundaries in the engineering task. But it is not universal because it is characterized by the engineering context within which their cross-site collaborations occur. Hence, dataspace is a general coordination mechanism utilized to manage cross-site dependencies through technology-mediated collaborations. Such collaborations and mutual awareness constitute means to harmonize tasks and information about it.[68]

However, just like workplace formation, dataspace utilization does not necessarily yield cross-site collaborations and GSE coordination. Engineers may not use communication media for collaboration if the dataspace is not challenged by software engineering problems that require cross-site collaborations – problems such as mutual unawareness and externally changing requirements. Therefore, how dataspace utilization is implicated in coordination requires an explanation of how the engineers use it to traverse different sites freely to address software engineering problems. They utilize dataspaces according to both organizational and personal communication preferences. Traversing space and usage of technology according to organizational and personal preferences signify appropriations of dataspace.

These appropriations are in response to the nature of the software engineering problems and the communication preference for addressing them. I call this *dataspace appropriation*. It occurs when someone uses a tool and operations in it to develop his own abilities.[69] Like the social meanings that identify workplaces as coordinative protocols, cross-site collaborations function among virtual team members through protocols. But different dataspace appropriations through different protocols imply different coordination processes enacted by developers to manage cross-site dependencies. These processes are ephemeral, but many of them are objectified continuously as coordination mechanisms in the utilized dataspace. In sum, space is implicated in GSE coordination because of engineering effectiveness; it is implicated by its utilization as a dataspace and by appropriating the dataspace to generate high dependencies.

To wit, virtual collaborations are generated with ICT resources and not necessarily by space. Space is rather the prospect or opportunity, says John Agnew, that organizations appropriate by relating it to ICT resources to create virtual workspaces. Appropriation of the GSE space to generate high dependencies implies that space is more relative to them than generative. High digital information dependencies in virtual workspace do not rely on space in the manner that high analog information dependencies rely on place. Space is, therefore, not an antecedent or generative mechanism but a relative mechanism for enhanced information management that achieves high dependencies across sites.

The upshot is that enhanced virtual, social, and economic processes in GSE at the product level leads to harmonization of task components. It suggests that spatial distribution of resources is a function of task variability at the product level. If the resources required for the engineering of the product are not harmonized, then it will be more difficult to develop. Therefore, it is proposed, additionally, that coordination efforts that aim to facilitate developers' information management capable of handling task difficulty are understood as increasing dependencies among product-level resources by harmonizing them with ICT in space. This implies that space, ICT, and related harmonized resources constitute the relative frame for inducing appropriate behaviors such as frequent interactions and efficient performance required to coordinate complex GSE (Figure 6.3).

From the analysis above, place and space correspond respectively to software component and product levels in the conceptualization of GSE coordination (Figure 6.4). On the one hand, component workplaces are formed as mechanisms

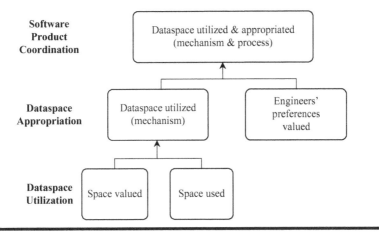

Figure 6.3   How space is implicated in GSE coordination.

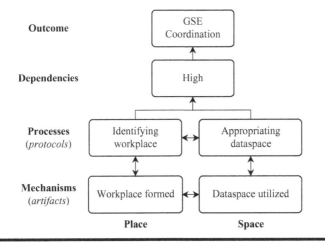

Figure 6.4   Place, space, and GSE coordination.

and then identified with communication processes for engineering efficiency. On the other, dataspaces are utilized as mechanisms and then appropriated through cross-site collaboration processes for engineering effectiveness. The two sets of mechanisms and processes are means of generating high task and information dependencies. In sum, it is proposed overall that GSE coordination results from managing dependencies by workplace formation and identification, and by dataspace utilization and appropriation.

This geographical perspective on GSE coordination incorporates the notable previous perspectives in the literature: ambidextrous coping whereby discipline in software engineering operations is combined with agility in handling changing external requirements;[70] shared team knowledge through organization design, work-based, technology-based, and social mechanisms;[71] team cognitive knowledge achieved by synchronized awareness of members' task statuses and actions;[72] time-separation cost reduction by appropriate choice of synchronous and asynchronous communication modes;[73] functional areas of expertise, product structure, process steps, and customization as alternative models;[74] and software architectures, plans, processes, and ad-hoc communications.[75] Interestingly, each of these perspectives exhibits an oversight of the combined phenomenological and general roles of place and space respectively, even though workplaces and dataspaces feature prominently in the structures and communication processes that shape coordination.

The oversight is mainly attributable to the assumption in these previous perspectives that geographical distance between places is a problem to be overcome through coordination. Because of this assumption, place has not been conceptualized positively as a significant variable of coordination theory in GSE. Researchers seem to be quite persuaded by the lure of the virtual,[76] and so they have distanced their models from placial or physical referents. As a result, they only slightly recognize these referents as problematic. For this reason, place, in particular, is essentially perceived as a phenomenon that is antagonistic to coordination theory. However, this chapter's proposal of placial and spatial perspective addresses these oversights because it explains how workplaces are formed and identified, and how dataspaces are utilized and appropriated to generate high dependencies that achieve coordination. The explanation has uncovered the underlying placial, spatial, and technological mechanisms and processes that shape the use of place, space, and ICT in GSE. Therefore, the proposal provides a complementary conceptualization of coordination.

The oversight is also attributable to scholars' predominant focus on logical structures and communication processes in accordance with Conway's law that assumes a strong correlation between organization structure and software product architecture. For instance, if this law is applied to the analysis in this research, then workplace and dataspace which were found to be coordination mechanisms, resulting from formation and identification of place, and from utilization and appropriation of space, would be attributed only to product structure requirements. However, the spatial perspective ascribes organization structure and communication processes to workplaces and dataspaces. This ascription contributes to the development of a more holistic explanation of coordination that accounts for geographical, technological, structural, developmental, and human factors in GSE coordination.

The spatial perspective shows that mechanisms and processes on the one hand, and place and space on the other, constitute a basic framework for analyzing GSE coordination (Figure 6.4). That is, the analyst who approaches GSE coordination from the perspective of place and space needs to analyze them as mechanisms and processes for better comprehension. The framework can be used to explain issues such as software integration efforts, formalization of procedures, communication and technology choices, and behavioral dynamics. Furthermore, it can be used to understand other effects of place and space such as mutual knowledge problems, engineers' inability to elicit spontaneous informal communication with colleagues in other sites, and meeting times. For example, it can serve as a framework for analyzing mutual knowledge problems and how knowledge integration can be used to build social capital. The model can also be studied in relation to particular organizational issues such as control, trust, conflicts, and ethnocentrisms.

Practically, this geographical perspective can guide a pro-active design of GSE for optimal coordination. Place and space are not easily made insignificant by ICT innovation. They remain as significant opportunities or constraints and must be exploited for the organizational design of GSE to optimize coordination. Therefore, when managers understand them in terms of their relationship with component and product structures, they can avoid disruptive adjustments and enhance continuities in high dependencies. This calls for place and space considerations to be incorporated by managers in task and related design decisions. To this end, considerations of task difficulty must incorporate place and space because they bear significantly on task performance. In globally distributed settings, these considerations will cause managers to align software components both to expertise and to place, especially when faced with changing customer requirements in software engineering.

Chapter 7

Paradox of Organization

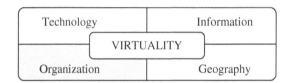

In organization theory, paradox refers to contradictory or conflicting yet inter-related factors that exist simultaneously. Paradox has been an influential framework for organizational research since its analytical power was espoused by Kim S. Cameron and Robert E. Quinn,[1] Marshall S. Poole and Andrew H. Van de Ven,[2] and Jennifer L. Sparr.[3] It is useful for analyzing complex, plural, divergent, and disruptive situations in organizations. Its philosophical and rhetorical origins are traceable to the Megaric philosophers of 400 BC who studied the well-known liar antinomy. Their famous example of paradox is this logical one: if a man says, "I always lie," how should we understand what he says? There are also rhetorical paradoxes which are statements that appear to be contradictory or absurd but are true. An example is in George Orwell's *Animal Farm*: "all animals are equal, but some are more equal than others." The theoretical tension in logical paradoxes and literary tensions in rhetorical paradoxes have led the way to the study of organizational paradox. In Poole and Van de Ven's admonition to scholars to embrace organizational paradoxes and use them to build theories, they wrote:

> An alternative strategy for theory building can be proposed: Look for theoretical tensions or oppositions and use them to stimulate the development of more encompassing theories. This strategy requires an exploration of the tradition of theoretical debate surrounding important issues, an identification of alternative or opposing theories or

explanations, and discovery of ways of relating, contraposing, or integrating them. The result will be theories less susceptible to the limitations of perspective which attend many middle range theories.[4]

On the whole, paradox has proven to be both an ontological and epistemological device for handling organizational tensions to generate theories. It confronts rather than ignores those tensions that reflect incompatible or inconsistent theories. The theories may appear inconsistent or incompatible, but that may be due to the underlying assumption, perspective, and explanatory principle. If theory development is a discourse, then to use paradox to develop global software engineering (GSE) coordination theory is to use it as a rhetorical device. For this reason, Poole and Van de Ven propose acceptance of paradoxes, spatial and temporal separation of paradoxes, and the introduction of new terms (literary devices) to resolve the paradoxes.

Wendy K. Smith and Marianne W. Lewis' most comprehensive review of the literature on paradox reveals paradoxes of organizing, learning, belonging, and performing.[5] Paradox of organizing manifests as tensions between competing arrangements and processes for the achievement of an organizational outcome (e.g. loose and tight coupling). Paradox of learning refers to tensions in the nature and pace of innovation (e.g. radical and incremental innovation). Belonging paradox reflects tensions of identity (e.g. the individual and the collective). Performing paradox is driven by tensions between demands of internal and external stakeholders (e.g. financial and social goals). Coordination of GSE is concerned with arrangements and processes for efficient production of software. Therefore, this paper applies the assumptions of organizing paradox to interpret the management of dependencies, conflicts, and uncertainties.

The theory adopted for this chapter's discussion is loose coupling by J. Douglas Orton and Karl E. Weick.[6] In a loosely coupled system, there are direct, more frequent, important, and strong links between resources that reflect the word coupling; and there are indirect, less frequent, less important, and weak links that reflect the word loose. By a method of loose coupling, a complex system can be decomposed into subunits in order to adapt it to external stimuli and to prevent negative effects from spreading through the system because they can be kept in one sub-unit. Organizations such as schools and hospitals, as well as multinational corporations, exhibit loose coupling as witnessed in the linkages between departments and international offices. Similar linkages are found in GSE organizations.

Loosely coupled resources and activities in an organization produce modularization, requisite variety, and behavioral discretion as direct effects.[7] Modularization manifests in GSE when loosely coupled resources exist as modular units or groups which are characterized by greater intra-dependencies at one site and less interdependencies across sites. This allows each modular unit to exercise greater concentration on its task component and reduces the spread of adverse stimuli among resources. Requisite variety is a state wherein the desirable variety in resources and behaviors matches the variety in the organization's internal and external environments; the basis for matching lies in managing the system's capacity to sense, register, and respond to its environment accurately. Behavioral discretion by global software engineers refers to the opportunity and capacity for them to exercise various autonomous actions in the face of environmental uncertainty. Audris Mockus and David M. Weiss[8] as well as David L. Parnas[9] argue that these direct effects are desirable conditions that should be consciously used as tactics for managing software engineering resources and activities. Since both requisite variety and behavioral discretion generalize variety in resources and behavior, these two tactics are jointly represented by the concept of variation. In sum, modularization and variation are the direct effects being applied to the analysis of GSE coordination in terms of organizational paradox.

At the same time, J. Douglas Orton and Karl E. Weick argue that loosely coupled resources constitute quite an unsatisfactory condition that should be moderated with unifying tactics such as leadership, focused attention, and shared values.[10] They call these tactics compensations for the direct effects of loose coupling. In the face of the potential presence of multiple and conflicting goals in loosely coupled systems, leadership provides centralized standards of direction and coordination while recognizing the value of increased discretion.[11] This can be achieved by using standard mechanisms such as time and targets to focus attention on controllable and essential behaviors. Shared values reflect socialization among workers which is achieved through cultural groupings, clan relations and mutual understanding over long periods of relationships, and it keeps loosely coupled systems from becoming anarchies. Compensations by both leadership and focused attention require centralized standardization, and so they are jointly represented by the concept of standardization. In sum, standardization and socialization are compensation for the direct effects being applied to the analysis of GSE coordination in this paper.

Loose coupling resonates with James D. Thompson's complex organization that is "an open system subject to criteria of rationality,"[12] implying that uncertainty is caused by an organization's openness to the external environment. It also resonates with J. Kenneth Benson's four principles of organizational dialectics: social construction processes; the totality of relationships or interconnections between organizational phenomena and structures; contradictions (rational and non-rational processes) involved in the social construction processes; and reconstruction of social arrangements.[13]

However, loose coupling is more suitable for this describing paradox because its tenets have stronger correlations with GSE coordination than these alternative ones. For instance, a tactic such as modularization of loosely coupled organizational resources correlates with modularization of software components and teams. Modularization has direct implications for dependencies, conflicts, and uncertainties among units, sites, and developers in GSE. According to James D. Herbsleb and Rebecca E. Grinter,[14] this is because the modular structure of a software product correlates with its organization structure (based on Conway's law).[15] Variation is another tactic in loosely coupled resources that correlates with variety in locations, cultures, experiences, and times that characterize GSE settings. These different variations also have direct implications for uncertainties and conflicts. For instance, the knowledge-intensiveness of software development produces the greatest expression of behavioral variation by developers who must respond to uncertainties with diverse innovations. Since modularization and variation are direct effects of loose coupling which are compensated for by standardization and socialization, it is logical that they also have direct implications for dependencies, uncertainties, and conflicts.

Moreover, loose coupling transcends J. Kenneth Benson's principles of organizational dialectics because it explains how loosely coupled resources are managed. His suggestions for managing organizational dialectics are summarized as "the processes through which actors carve out and stabilize a sphere of rationality and those through which such rationalized spheres dissolve."[16] But his suggestions are quite generic compared with those of loose coupling that specify particular tactics such as modularization, standardization, variation, and socialization.

Orton and Weick suggest that applying direct effects or their compensations to the analysis of loosely coupled systems leads to uni-dimensional interpretations that are limited. In order to provide an integrated conceptualization of GSE coordination, both direct effects and the compensations are interpreted as a paradox of tactical tensions.

To learn about the paradoxical interpretation of coordination, consider a collocated software development system (site *A*). Consider also the outsourcing of aspects of software development operations to a remote location (site *B*) (Figure 7.1). By virtue of this distribution, site *A*'s operations may still be more protected from direct environmental influences by its managerial and institutional levels than site *B*'s.[17] However, so long as operations in both sites are interdependent, site *A* is essentially more open to the environment than before since environmental stimuli affecting site *B* would inevitably affect it. This undermines the buffers at the managerial and strategic levels which the organization enjoyed before. Therefore, the organizing logic of operations changes from closed to partly closed and partly open.[18]

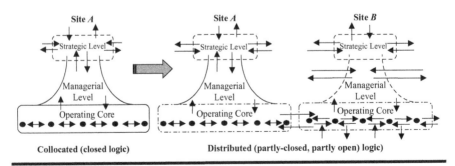

Figure 7.1 Collocated and distributed software engineering.

Note: The arrows signify information, human, and material flows.

These direct influences mediate the relationships between software engineers, information, and technologies; and they reflect both desirable and problematic effects in terms of their management to achieve coordination. Given this backdrop, the review of coordination is grounded on the following descriptive framework: organization of GSE is characterized by paradox of loose coupling; and coordination is achieved by the process of managing dependencies, uncertainties, and conflicts among the resources by exploiting the desirable direct effects and compensating for them.

In the face of distributed GSE resources, organizations modularize software components in such a manner that task dependencies are increased at particular sites and reduced across sites leading to their enhancement.[19] This is because *modularization* achieves more efficient and effective logical workflow in software development.[20] Modularization in GSE reduces the costs engendered by the distribution, complexity, and knowledge-intensiveness of software development. The costs are epitomized by communication delays,[21] communication breakdowns,[22] and knowledge sharing or transfer problems.[23] For example, Christof Ebert and Philip De Neve have demonstrated that task collocation by Alcatel's Switching and Routing business division achieved an efficiency improvement of over 50% during initial validation activities.[24]

One of the main ways by which the need for physical proximity between developers underscores modularization is the arrangement of office spaces in such a manner that they can more easily engage in verbal and spontaneous interactions.[25] Through these interactions, site dependencies and uncertainties are better managed to achieve coordination because developers can gain greater workflow awareness.[26] But the same arrangement undermines cross-site dependencies and uncertainties because of reduced cross-site awareness and interactions.[27]

Furthermore, modularization of task organization reduces the spread of adverse stimuli among the resources which constitute the software development system.[28] The low spread of stimuli enhances the management of environmental uncertainties which confront developers at one site. Without modularization, the development system is susceptible to rapid spread of environmental uncertainties in the software development system, even if there is loose coupling between resources and activities.[29]

At the same time, software development organizations aim to ensure that dependencies between sites are enhanced and related uncertainties reduced in order to compensate for the direct effects of modularization. This aim is in accordance with rational organization through centralization or *standardization*.[30] Typical examples of structural mechanisms that ensure the prevalence of some standardized behaviors among developers are formal methods,[31] documentation requirements,[32] information and communication technology (ICT) support,[33] and architectures, plans, and processes.[34] These standardized mechanisms and behaviors are meant to reduce cross-site uncertainties and dependencies because they hold the organization together to minimize uncertainties that may degenerate into anarchy. These standardized mechanisms are designed to be generic, with allowance for their adaptation to local situations.[35] For example, in their study at ComSoft and Witel, Suprateek Sahay and colleagues found that standardization was used for long distance control, although it was met with local resistance in both cases.[36]

Alongside modularization, GSE suggests a preference for requisite variety in resources to match the variety in environmental uncertainties.[37] This is in accordance with the law of requisite variety which states that "the variety within a system must be at least as great as the environmental variety against which it is attempting to regulate itself."[38] For instance, the decision by a GSE organization to draw software development expertise from diverse locations around the world already implies a consummation of locational requisite variety to achieve close-to-market development gains.[39] Moreover, the systems development literature shows a positive relationship between developers' autonomy, discretion, and performance when faced with volatile environmental uncertainties.[40] Behavioral discretion is a safe management tactic to exploit for GSE coordination because as task and environmental uncertainties are varied and emergent, so must there be autonomy for developers to gather information and make innovative and spontaneous decisions. For example, Likoebe M. Maruping and colleagues found that when developers exercise autonomy in agile development, it enables them to accurately sense, register, and respond to last-minute changing requirements.[41] Therefore, variation reduces environmental uncertainties.

At the same time, it can cause both site and cross-site misunderstandings between developers and increase interpersonal conflicts between them. This is especially the case when there are dissimilar individual and site-specific cultures and expectations in a GSE team. For example, David J. Armstrong and Paul Cole found from their study of an American company that "misunderstandings developed on the basis of different assumptions about the tasks and assignments,"[42] leading to increases in both site and cross-site conflicts between developers.

The compensation for the undesirable direct effects of modularization and variation is *socialization* among software developers. According to control theory applied in agile software development contexts,[43] socialization is a very appropriate mechanism for achieving trust and shared values. For example, Iván Alfaro's study shows that facilitating and mentoring software engineers enhance their interpersonal interactions and relationships; and the enhanced relationships engender a positive relationship between national diversity and team performance.[44] Ilan Oshri and colleagues also show in their research that face-to-face meetings are indeed important in creating and facilitating interpersonal ties.[45] Likewise, Angelika Zimmermann's review of interpersonal relations in virtual software teams points to high social capital benefits they gain from the relations.[46] These ideas corroborate Nils B. Moe and Darja Šmite's finding that poor socialization is one of the major causes of a lack of trust in global virtual teams.[47] Socialization and resultant shared values produce a strong sense of community,[48] and fulfill the requirement for organic information generation and processing.[49] Organic information processing is required for complex research and development tasks such as GSE.

Concerning the relationship between socialization and dependencies and uncertainties, Pamela J. Hinds and Cathleen McGrath found no support for the argument that social ties facilitate cross-site coordination ease.[50] However, their research was an examination of social ties enabled by computer-based social networks. This is an instance of work that requires organic information processing to achieve coordination but is rather saturated with mechanical information processing. Michael L. Tushman argues for the necessity of organic information processing in the form of verbal, informal, and spontaneous communications between software developers since software development requires this type of information processing.[51] Such communications are hard to achieve through computer-based social networks. Julia Kotlarsky and Ilan Oshri confirm Michael Tushman's argument within the context of GSE because they found that social ties between distributed developers, underpinned by face-to-face communications, rapport, and trust, actually facilitate coordination and collaboration.[52] Therefore, optimized socialization, that is social ties underpinned by face-to-face communications, enhances dependencies and reduces uncertainties across sites.

Just over a decade ago, James D. Herbsleb wrote about the future of sociotechnical coordination in global software engineering, expressing the "need for a systematic understanding of what drives the need to coordinate and effective mechanisms for bringing it about."[53] By interpreting the paradoxical forces undergirding coordination, this chapter has described the trade-offs between the direct effects and compensations of loose coupling among GSE resources. The chapter set out with the following premises: that paradoxical interpretations are required for an integrated conceptualization of GSE coordination. It has shown that organizations do not only focus on management of generic constructs to achieve coordination. Additionally, they focus on the management of specific constructs that are pertinent to GSE challenges. Their focus on pertinent constructs leads us to shift from a generic to both a germane and comprehensive knowledge of the paradox of GSE coordination.

It is proposed here that particulars of dependencies such as cross-site and site; of uncertainties such as site, cross-site, and environmental; and of conflicts such as site, cross-site, intercultural, and interpersonal are pertinent constructs. In previous research, these pertinent constructs have been overlooked because researchers have focused largely on functional explanations which seek functional relationships between coordination and other GSE variables. This is witnessed, for instance, in Rajiv Sabherwal's explanation of the evolution of coordination when mechanisms such as plans, mutual adjustments, and standards are applied to manage outsourced projects.[54] It is also witnessed in explanations of how coordination relates to other organizational variables such as systems architecture,[55] temporal distance,[56] and developers' knowledge.[57]

The focus on functional explanations at the expense of essential ones has masked the pertinent constructs that have been shown in this paper to be essential bases of coordination have been masked. Hence, knowledge of the paradox of GSE coordination in previous research has been inadequate. Because these pertinent constructs reflect key parameters such as distance, environment, task structure, developers, and culture, they render this proposed epistemology of the paradox of GSE coordination more pertinent. They also enable greater exercise of research and managerial control of resources and activities.

This chapter also presents a more comprehensive knowledge of the paradox of GSE coordination than previously known by identifying multiple pertinent constructs. Previous explanations have largely paid attention to the three generic constructs, and so our understanding has been less comprehensive. Again, the prevalence of this limited or generic attention is traceable to explanations that are replete with functional relationships between coordination and other organizational variables. This chapter rather demonstrates that management of multiple pertinent constructs of dependencies, uncertainties, and conflicts provides us with comprehensive knowledge of the paradox of GSE coordination.

Although this paper has primarily sought to provide a conceptualization through a review, it also offers important practical guidelines. The proposed pertinent constructs of the paradox of GSE coordination generate specific and challenging conditions for practice. First, when practitioners come to terms with this pertinent and comprehensive knowledge of coordination, it will lead them to focus on the pertinent constructs and consciously manipulate them to achieve coordination. This is because knowledge of multiple pertinent constructs, being the essentials and bases of sound coordination, present managers with tools for gaining more control over the coordination process.

Second, this pertinent and comprehensive understanding of the paradox of GSE coordination draws managers' attention to the ambivalent character of the global distribution of resources and activities for their coordination. Global distribution is characterized by ambivalence because it can be equally the source and resolution of coordination challenges. Managers can learn from this ambivalent character in order not to treat distance as only a coordination problem to be overcome, but to treat it also as a resource for resolving contradictions that occur during GSE coordination. The broader understanding of the pertinent constructs as the bases of coordination provide managers with a more robust instrument to look beyond particular direct effects and compensations of global distribution. The instrument enables them to take both systemic and dialectical approaches to coordination at the same time.

Third, the co-existence of multiple pertinent constructs (some of them being opposed to others) suggests that coordination is an unending and dynamic challenge. These constructs point to the possible existence of negative effects of global distribution at all times. They require continuous attention by managers to prevent them from undermining their coordination efforts. Continuous attention to the oppositions and trade-offs is especially required given that the same intervention to compensate for a negative direct effect may generate another negative effect. This chapter informs both managers and developers of the details of these trade-offs in order that they will practice coordination by tolerating significant levels of the negative effects.

The description of coordination in terms of the paradox of organization has brought forth pertinent constructs, and points to several implications for research and development of new theory. Because coordination has implications for both software engineering and ICT, the paradoxical conceptualization has several theoretical, analytical, and methodological possibilities which are discussed below.

First, this proposed knowledge of the paradox of GSE coordination should become a major focus of theoretical explanations in future research. The pertinent constructs discussed in this chapter should encourage researchers to undertake a more in-depth analysis of this reality. For example, what would be the character of coordination if one set of pertinent constructs (e.g. interpersonal conflicts and environmental uncertainties) dominate and are given management priority over the others: how and why? The greater depth of analysis afforded by this knowledge of the paradox of GSE coordination will help researchers to obtain more accurate explanations of different portfolios of coordination. There are precedents of researchers' pursuits of in-depth explanations of organizational dynamics in software development research. For example, deeper analysis and explanations of software engineering control have been provided by researchers through studies of how control is exercised differently in different phases of product development,[58] and for tackling different product development challenges.[59] Scholars who undertake in-depth analysis will have their attention drawn to on-going processes of software development, leading to explanations of how and why management of the pertinent constructs are enacted by organizations.

Second, the deeper and comprehensive understanding of the paradox of GSE coordination suggests that future researchers who relate coordination to other organizational variables should uphold its paradoxical nature. This will lead to two main analytical benefits. In the first place, it will foster comprehensive analysis because it induces the analyst to account for all coordination resources and mechanisms in a GSE system, and to arrive at richer and more accurate conceptualizations. In the second, it will foster dialectical analysis by enabling the researcher to think simultaneously about contradictory pertinent constructs of coordination. Researchers can use paradox constructively to build theory if they clarify levels of analyses.[60] The knowledge of the paradox of GSE coordination has disclosed multiple pertinent constructs and how they are managed by organizations to achieve coordination. Thus, the paradox of GSE coordination provides a framework for both comprehensive and dialectical analyses in future research.

Third, the paradox of GSE coordination as a dynamic process traceable to management of the pertinent constructs is tentative and requires further testing. If there are different possible portfolios of coordination that can be explained, then functional explanations of the impact of coordination on other organizational issues such as performance, innovation, culture, knowledge, and transformation will address this testing requirement. Because the proposed epistemology of the paradox of coordination is pertinent and more comprehensive, researchers can apply it as an approach to explain other organizational issues. Prasert Kanawattanachai and Youngjin Yoo's explanation of how virtual team performance is shaped by knowledge coordination is a typical example.[61] In short, while the pertinent constructs provide clearer understanding of the paradox of GSE coordination, further researcher research on how it impacts other organizational variables in software development will explain its functions.

Fourth, since GSE is made possible by ICT innovation and deployment, further studies of the impact of current phenomena such as virtualization and digitization on the pertinent constructs are in order. A major aspect of GSE is virtual teamwork induced by ICT innovation and deployment, and so studies of the relationship between virtualization and these pertinent constructs promise new insights. Precedents of such studies can be gleaned from previous explanations of coordination in terms of electronic networks,[62] tool support,[63] collaboration technology embeddedness,[64] and technology adaptation.[65] Researchers can draw inspiration from this ongoing stream of research and from continuing ICT innovation to further explore how and why other aspects of virtualization such as simulation and representation[66] shape the pertinent constructs.

Fifth, one of the factors that underscores pertinent constructs is the spatial distribution of and the distance between software resources: people, technology, and information. It is the spatial factor that has enabled our understanding of dependencies, uncertainties, and conflicts in terms of site and cross-site constructs. It is also the factor that catalyzes interpersonal conflicts and environmental uncertainties. Yet previous GSE coordination explanations treat it as a contextual rather than a theoretical issue. This is in spite of the fact that GSE researchers have provided scattered accounts of opportunities and limitations of the spatial factor in previous publications. The exclusive attention to people, technology, and information resources lead researchers to pose coordination questions of what, how, and why at the expense of where. Consequently, existing explanations of how spatial distribution relates with these software resources to achieve coordination have not yet been provided. This chapter's disclosure of these pertinent constructs represents direct attention to the spatial factor as a theoretical issue that should be explored in future research.

The paradox of GSE coordination described as managing direct and inverse relationships between multiple dimensions of dependencies, uncertainties, and conflicts is generic compared with previous ones. This is due to the more integrated and nuanced approach that has led to the combination of dependencies, uncertainties, and conflicts in this conceptualization. At the same time, the paradox of GSE coordination being proposed is accurate compared with previous ones because of the interpretations which have unearthed the multiple dimensions and their inverse relationships in the pertinent constructs. However, the proposition is not as simple as it is general and accurate, and this limitation should be expected if it is understood in terms of Warren Thorngate's postulate of commensurate complexity. According to this postulate, a theory of social behavior, epitomized by GSE coordination, cannot be general, accurate, and simple at the same time. It is admitted that the proposed conceptualization lacks simplicity because it ascribes coordination behavior to several factors such as dependencies, uncertainties, distance, knowledge-intensiveness of software development, and dialectics. For this reason, a charge by critics that the proposed conceptualization sacrifices economy in explanations may be in order. However, the explanations have not been informed by intentional over-determination of factors that are really necessary to conceptualize coordination. Each of the factors included in the explanations are significant determinants of the paradoxical and integrated dynamics of coordination.

This chapter has presented an extensive review of the literature on GSE coordination, pointing to a fragmented understanding of its nature and to paradoxical relationships between its constructs. To address the problem of fragmentation, the chapter has undertaken a paradoxical interpretation of the constructs to contribute new knowledge of the basic construct and integrated structure of GSE coordination. The paradox of GSE coordination (the basic construct and integrated structure) is explained as managing direct and inverse relationships between multiple dimensions of dependencies, uncertainties, and conflicts. The conceptualization points to this book's contribution of an integrated epistemology and a specific ontology of GSE coordination.

Chapter 8

Virtuality of Coordination

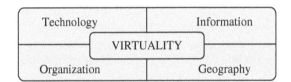

Here is a recapitulation of the main arguments underlying the motivation behind this book. Virtuality enables representation and simulation of software engineering activities that permit new globally distributed coordination practices founded on information and communication technology (ICT). Virtuality is a description of a new form of organization that is characterized by geographic dispersion, electronic dependence, structural dynamism, and national diversity that may hinder team effectiveness.[1] These characteristics present theoretical and practical challenges for coordination which is the management of dependencies, uncertainties, and conflicts in organizational activities.

Global software engineering (GSE) coordination has attracted considerable research attention since the end of the last century, leading to coordination theories which reflect traditional instead of virtual organization. In traditional organization, coordination is basically informed by rationality and indeterminacy logics. Rationality refers to the "extent to which a series of actions is organized in such a way as to lead to predetermined goals with maximum efficiency";[2] indeterminacy refers to "interconnections among dependent parts [that] are somewhat less constrained, allowing for flexibility of response."[3]

On the whole, existing coordination theories overlook virtuality which is a distinct logic transcending the contradictory and traditional logics of rationality and indeterminacy. Hence, our understanding of technological explanations of coordination informed by virtual logic has been confined within these logics. There is therefore the need for a GSE coordination theory that incorporates but transcends existing theories. The theory should account for virtuality as it is implicated in the technology, organization, information, and geography of coordination. This chapter presents the integration of the four coordination perspectives with the virtual approach.

To begin with, existing coordination theory in the general organizational and the particular GSE renditions have to be recast in terms of rationality and indeterminacy logics. These logics were discussed in Chapter 3, mostly in their definitive terms because the aim there was to justify the distinctiveness of virtual logic in organization. Their applicative terms were hinted at in the justification in these words: "Virtuality as an essential logic of organization is first understood in terms of the peculiar task coordination challenges it addresses, just as the logics of rationality and indeterminacy are understood by their peculiar task coordination challenges they address." The focus there was on how task coordination challenges justify the virtual logic, but not on how task coordination solutions or theories provide that justification. Moreover, the relationship between rationality and coordination as well as between indeterminacy and coordination were not discussed. The discussions there on the logics of rationality and indeterminacy were in relation to general organization but not to the particular issue of coordination. Indeed, the purpose of the entire chapter was to argue for virtuality as an essential and distinctive organizational logic, underpinning a later argumentation for virtuality as an analytical approach to GSE coordination. In short, the discussions there did not provide adequate elaborations of relationships between virtuality and coordination; and this book has so far not undertaken that elaboration.

Given that this relationship is explained by appraisal and integration of the four coordination perspectives, the explanation must provide justifiable reasons for how and why we should depart from coordination theory informed by rationality and indeterminacy toward the proposed coordination informed by virtuality. The explanation must also provide justifiable reasons for how and why coordination theory informed by virtuality constitutes an integration of the existing four perspectives previously cast in terms of rationality and indeterminacy logics. In order to satisfy these conditions, four main assumptions are used.

The first is that the theoretical task of integrating the four coordination perspectives is understood as a task of synthesis. Hence, a synthesizing analytical framework is required. Second, virtuality transcends rationality and indeterminacy, yet it includes them because they are fundamental realities in human affairs. The discussions of the materiality of ICT (Chapter 4) and the exploitation of geography (Chapter 6) also point to the realities of ICT, places, and spaces. The third is that the paradox of organization (Chapter 7) informs us that paradoxes in GSE coordination theory and practice are offshoots of the two countervailing forces; that is, coordination informed by rationality and coordination informed by indeterminacy. Hence, a framework for the dialectical analysis of paradoxes is required. The virtual is defined by the real, and so a framework for analysis of the relations and translations between reality and virtuality is required. Fourth, virtual is abstract, and the integration of the four coordination perspectives with a virtual approach is an abstraction. Hence, the analysis here requires an interpretive framework that enables the development of an abstract theory of coordination.

These four assumptions suggest to us Georg W. F. Hegel's theory of dialectical idealism.[4] Dialectics, in general, is a philosophical theory which explains nature, history, and human behavior in terms of contradictions.[5] It assumes that the world is characterized by contradictory forces or propositions which are the outcome of an absolute idea – that idea is not obvious because it is abstract. In dialectical reasoning, the analyst identifies the known countervailing forces or propositions in terms of thesis and antithesis. The thesis is a proposed theory about a phenomenon in the world (usually the earliest one); the antithesis is another theory proposed to negate the thesis (usually a later one). Thus, coordination theories informed by rationality are the thesis, and those informed by indeterminacy are the antithesis (rationality theories also antedate indeterminacy ones in organizational studies literature). Using dialectics, the analyst then seeks to resolve the countervailing theories by proposing a synthesis which also negates the antithesis. The synthesis emerges as the absolute idea which is the higher and primary explanation for the phenomenon (which exhibit the thesis and antithesis) (Figure 8.1).

The Greek philosopher Heraclitus explained dialectics with the example of a bow. The bow generates force to propel an arrow when the string and wood are drawn in opposite directions. Thus, the string and the wood oppose yet complement each other as thesis and antithesis (push and pull). The propelling force generated is the synthesis which provides a higher and primary explanation of the bow. That is, each of the string and wood is a lower and secondary explanation of the bow; the generated force provides the higher and primary explanation which transcends but incorporates the string and wood. The discussion and example show

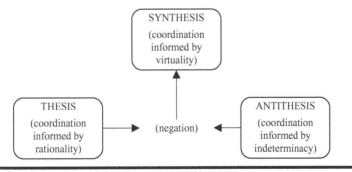

Figure 8.1   Dialectics.

that negation and contradiction in dialectics do not necessarily mean the denial or absolute opposition of the countervailing theories. It essentially means two theories may imply one another (e.g. inside and outside), or they are conceptually interrelated (e.g. general and particular), or they are intimately interrelated (e.g. quality and quantity).

Thus, dialectical idealism in particular assumes a "world spirit" or idea, a system of abstract universals that explains things and events in nature and history. The idea is neither spatial nor temporal; it is abstract, but finds its concrete expression in nature which is material. Therefore, absolute idealism interprets nature and history in terms of the primacy of the idea. In Hegel's original formulation, the absolute idea is the synthesized one. It is the present reality that reflects dialectical development through negation of an antithesis which also negated a thesis. Even though the development process includes the thesis and antitheses as past or observable circumstances, Hegel "finally reduces the causal efficacy of the past in its influence on the present to the present's act of emerging by going beyond, but also retaining and conforming to, elements of its developmental past."[6]

Karl Marx and Friedrich Engels, in their alternative system called dialectical materialism, critique Hegel's dialectical idealism as too abstract.[7] In dialectical materialism, Marx and Engels rather interpret history in terms of the primacy of processes and practices of human labor. Hence, Hegel's absolute idea does not determine dialectics; rather, dialectics are necessary outcomes of the class struggle between the working class and owners of production as they seek to satisfy material needs in industrial society. However, dialectical materialism should be understood as a practical explanation quite distinct from dialectical idealism, which focuses on abstract explanation. This book draws upon dialectical idealism because it seeks to develop an abstract explanation of GSE coordination.

Dialectical idealism is argued further to be the most appropriate framework for this explanation because of the following reasons. First, it recognizes countervailing propositions as this book does, and facilitates the analysis to integrate and transcend them. Therefore, it maintains these propositions in the synthesized theory, and facilitates analysis of virtual organization of GSE without abolishing the traditional version. Second, it problematizes the countervailing propositions in order to depart from them, leading to the pursuit of a higher and broader explanation of GSE coordination. Its appropriateness as an analytical framework is further demonstrated in Rudy Hirschheim and Heinz Klein's emancipatory principles methodology[8] and Gro Bjerknes' theory of dialectical reflection.[9] It has also been used for theorization of information systems implementation in terms of collective minding,[10] technology and process changes,[11] hermeneutics,[12] innovation,[13] and social inertia.[14] Interestingly, it has not been applied to synthesize GSE coordination context in spite of the countervailing thesis of coordination theory informed by rationality and the antithesis of coordination theory informed by indeterminacy.

Dialectics and GSE Coordination

The second and most important aim of this book is to integrate the four perspectives of coordination in order to propose a new explanation of GSE coordination practice. The integration exercise must show how and why the virtual logic premise and the four perspectives constitute an improved theory of technology coordination. The new explanation must point to a more holistic understanding of how and why ICT is the essence of GSE coordination.

We have now come to the integration exercise where dialectical idealism is used to explain GSE coordination as a synthesis of the thesis (which is coordination informed by rationality – summarized as coordination by plans) and the antithesis (which is coordination informed by indeterminacy – summarized as coordination by mutual adjustments). Thus, the analysis begins with a discussion of the thesis to show how the four coordination perspectives are represented there. This is followed with a discussion of the antithesis with a similar objective. The discussions culminate in this synthesis: the absolute idea of GSE coordination informed by virtuality and transcends, but includes rationality and indeterminacy.

Coordination by plans is the establishment of schedules for interdependent units and to govern their actions.[15] Organizations that practice collocated software engineering aim to ensure that dependencies between sites are enhanced and related uncertainties reduced. This aim, in accordance with rational organization, is achieved through centralized direction and standardization.[16] To standardize organization is to conform means, methods, and behaviors to some agreed level of efficiency. The idea is to enable easy optimization of alternative actions for goal implementation by application of mathematical and linear methods to systematic selection of behaviors. Standardization is necessary to hold the organization together and minimize uncertainties without necessarily supplanting behavioral discretion and task modularization.[17]

In the context of GSE, one way by which organizations achieve standardization is by adopting particular structural mechanisms to govern dependent relations between software engineers and the software task components. Architectures, plans, and processes are typical structural mechanisms that ensure standardized behaviors among developers.[18] Thus, in their case study of software engineering coordination in an ICT organization dispersed in three locations in Finland, Päivi Ovaska and colleagues found that coordination of dependencies is achieved through architecture descriptions and well-defined interfaces between components.[19] Similarly, in their study of component-based software engineering at TCS and LeCroy which were distributed between India, Switzerland, and the United States, Julia Kotlarsky and colleagues also found that standardization and centralization of engineering tools across sites led to effective component reuse through cross-site coordination.[20] Laurie Kirsch and colleagues,[21] as well as Likoebe Maruping and colleagues,[22] also reported from their empirical studies that process and outcome control of large information systems projects is achieved by the use of standard operating procedures or methods (such as coding standards) and performance evaluation standards. Similarly, Gwanhoo Lee and colleagues "suggest that global software teams should standardize software processes to help overcome difficulties resulting from team dispersion across multiple boundaries."[23] Ilan Oshri and colleagues' study of knowledge transfer among globally distributed software team members also points to how the use of codified transactive memories including standardization of templates and methodologies contribute to expertise coordination across sites.

A second way organizations achieve standardization in fulfillment of coordination by plans is by identifying major or overarching resources in GSE configurations such as space, time, and technology for high level control. For example, James D. Herbsleb and Deependra Moitra argue that the scheduling of engineering activities is an instance of using time as a standard mechanism that limits engineers' discretion in the choice and performance of activities.[24] In the context of collocated software engineering, Frederick Brooks also argues that while developers at the lower level may be permitted to exercise behavioral discretion in the choice of interaction media, such discretion is limited by the standardized controlling measures inscribed into technology.[25] This is a classic argument which is extensible to the GSE context because it has low geographical and temporal connotations.

But how are organization, technology, information, and geography implicated in the classical thesis of coordination by plans (informed by rationality) that has the objective of standardization? If the purpose of standardization is to hold the organization together by minimizing uncertainties and rationalizing dependencies according to rational logic, then organization and information predominate at the expense of technology and geography in the rational approach to theorization of coordination. Standardization itself assumes that organizational paradox is a problem in GSE due to uncertainties generated by changing software requirements from customers, geographical distribution, socio-cultural differences among software engineers, and the complexity of the software engineering task. Therefore, standardization is the counter tactic to suppress organizational paradox by rationalizing dependencies through the use of architectures, standard operating procedures or methods, and engineering schedules and templates, as well as codified transactive memories for knowledge transfer among software engineers. But, the sole application of rational logic to GSE coordination is theoretically and practically limited because it cannot address uncertainties generated due to software requirements by customers, geographical distribution, socio-cultural differences among software engineers, and the complexity of the software engineering task. Standardization is imbued with intentions to suppress the paradox of organization, leading to the management of information, but paradoxes and uncertainties remain in GSE. This implies that paradox of organization and management of information are weakly implicated in the classical thesis of coordination by plans (informed by rationality) that has the objective of standardization.

Related to this is that exploitation of geography and materiality of technology are also weakly implicated in the rationality approach. Geography is problematized rather than exploited in this approach. Thus, Gwanhoo Lee and colleagues recommend that GSE organizations should "standardize software processes to help overcome difficulties resulting from team dispersion across multiple boundaries."[26] Because of researchers' persistent adherence to rational logic in analyzing GSE coordination, exploitation rather than problematization of geography is yet to be given theoretical attention. So standardization is understood as a tactic that is employed to counter the problem of geographical distribution of software engineering resources and activities. But that is adopting the rational approach to its logical end. This understanding points to the Newtonian view of place and space whereby place is given priority over space and researchers argue that coordination is achieved through placement of resources. The Leibnizian view which prioritizes space over place and emphasizes possibility spaces enabled by ICTs is overlooked in GSE coordination approached with rationality.

Similarly, technology is given an instrumental rather than a material role in the standardization tactic. Protocols in ICTs that constitute standards for information exchange and management among software engineers are merely incidental to coordination because researchers interpret them as instruments for rationalization and formalization of information exchanges. Even though in practice GSE organizations employ varied technologies, communication modes, and information formats, the rational approach that drives standardization leads to researchers' oversight of these variations and their materiality in coordination success. This instrumental interpretation of ICTs reflects researchers' skewed perception and analysis of them as mechanical media, even though they are predominantly electrical media in reality and in practice. Interpretation of ICTs as mechanical media, according to standardization in the rational approach, is clearly witnessed in the low theorization of how organizations combine ICTs with geography to achieve coordination.

Recall that the thesis of our dialectical interpretation of GSE coordination is coordination by plans (informed by rationality), the antithesis is coordination by mutual adjustments (informed by indeterminacy). Coordination by mutual adjustments is the use of varied and flexible behaviors such as verbal, ad hoc, and informal communications among workers in order to address variable and unpredictable situations.[27] In the face of the global distribution of resources and activities leading to varied and emergent uncertainties, the literature on coordination by mutual adjustments proposes modularization, variation, and socialization as negations of standardization. They are interpreted dialectically as negations because they are supposedly non-rational tactics.

To effectively coordinate software engineering resources and activities, organizations modularize software components in such a manner that task dependencies are increased at particular sites and reduced across sites leading to their enhancement. According to David L. Parnas, software modularization achieves more efficient and effective logical workflow in software engineering.[28] Modularization reduces uncertainties at particular sites and increases them across sites because modularization also achieves frequent, verbal ad hoc, and informal interactions in one location and avoids inhibited communication and coordination between software engineers at different sites.[29] Modularization has both logical and physical dimensions which reflect task components and geographical location respectively (see Chapter 7).

Variations in organizational resources and activities constitute a tenet of coordination by mutual adjustments – it is presented as a negation of standardization. The basis of the argument is that a variety of organizational resources and activities is required to match the variety in task uncertainties (the law of requisite variety). The assumption is that indeterminacy, which is epitomized by uncertainty, is an indispensable aspect of organizational reality. Indeed, GSE entails huge doses of indeterminacy because of uncertainties generated by changing software requirements, by geographical distribution, by socio-cultural differences among software engineers, and by the complexity of the software engineering task. The proposition of variation as a coordination tactic is drawn from how software engineers are capable of exercising behavioral discretion through improvisation and dexterity to address such uncertainties. They exercise variation by employing various ICTs, communication modes, and information formats according to their various preferences and peculiar challenges (see Chapter 5). The software engineering literature shows a positive relationship between engineers' autonomy and performance, especially in the face of uncertainties from changing customer requirements.[30]

Here are some of the notable arguments. Autonomy, as exemplified in agile methodology, is necessary for engineers' accurate sensing, registering, and responding to the environment in the face of last-minute changing requirements.[31] Similarly, knowing-in-practice[32] by software developers as well as their improvisation[33] and ambidexterity through flexibility and agility[34] are employed to cope with environmental uncertainties. Gwahnoo Lee and colleagues suggest that "software teams should build process agility to cope with requirement changes and that the benefit of agility is greater when user requirements change faster."[35] To cope with uncertainties such as missed deadlines and failures, Alberto Espinosa and colleagues propose shared knowledge of the team and task as the useful coordination processes.[36] When uncertainties are in the form of task contingencies, Deepa Mani and colleagues found that information sharing across sites is effective for task performance.[37] Therefore, variation as an antithesis of standardization is a central tenet of GSE coordination by mutual adjustments, suitable for an indeterminacy approach to coordination theory.

In spite of the benefits of variation, it is also attended by conflicts among software engineers at the same time. The very exercise of discretionary behavior to address emergent and varied uncertainties, coupled with socio-cultural differences among engineers at different sites, can generate interpersonal conflicts. Discretionary behaviors may reflect task uncertainties or dissimilar individual and site-specific cultures and expectations in a GSE team. While they are desired variations, they pose problems of interpersonal conflicts. David J. Armstrong and Paul Cole's study of the American company points to such interpersonal conflicts. Their work is corroborated by Pamela Hinds and Diane Bailey's theoretical study of conflict in globally distributed teams.

For these reasons, the GSE coordination literature proposes socialization as a process for resolving conflicts arising from variation. This proposition is in harmony with the practice of socialization as a coordination process for conflict resolution in software organizations around the world. A notable example is the mentoring of less experienced software engineers by more experienced engineers, as demonstrated in Iván Alfaro's research.[38] Another example presented by Ilan Oshri is face-to-face meetings that may involve long-distance travel by engineers between sites.[39] According to Angelika Zimmerman's review of interpersonal relations in virtual software teams, software engineers gain high social capital benefits. Poor or inadequate socialization, argues Nils B. Moe and Darja Šmite, is one of the major causes of mistrust in global virtual teams.[40] Sound socialization promotes shared values as well as a strong sense of community among software engineers.[41] According to Michael Tushman,[42] socialization is an important requirement for organic information generation and processing.

Having established that modularization, variation, and socialization reflect the indeterminate approach to GSE coordination, how can we understand how organization, technology, information, and geography are implicated in this approach? Analysis of the arguments for modularization, variation, and socialization suggests that organization and information predominate at the expense of technology and geography in the indeterminacy approach to theorization of coordination. This is similar to how these four perspectives are implicated in the rationality approach. But the difference here is that organization predominates in the sense that paradox is tolerated rather than suppressed. The three arguments (modularization, variation, socialization) assume that organizational paradox is real and acceptable so long as the human capital of software engineers can be harnessed to manage them. Thus, uncertainties generated from changing software requirements by customers, geographical distribution, socio-cultural differences among software engineers, and the complexity of the software engineering task are tolerable because they are manageable. And indeed, the practice of coordination by mutual adjustments, as reported in the literature (and as this book also illustrates in Chapter 9), testifies that sound information management with modularization, variation, and socialization occurs. Global software teams' exercise of these three tactics is imbued with intentions to absorb indeterminate circumstances that, combined with rational ones, underscore the paradox of organization. In sum, the paradox of organization and management of information are strongly implicated in the antithesis of coordination by mutual adjustments (informed by indeterminacy).

In spite of the strength of organizational paradox and information management in the theorization of GSE coordination by mutual adjustments, exploitation of geography and materiality of technology are strangely weak. The strangeness is due to the fact that GSE coordination practice strongly includes the exploitation of geography and materiality of technology. Geography is still problematized rather than exploited in this indeterminacy approach to theorization. On the ground, the placement of software components at particular sites (physical modularization) to increase site dependencies and reduce cross-site ones according to Conway's law is a practice, but it has not been adequately conceptualized in terms of exploitation of geography. This low conceptualization remains because of low integration of geography, technology, information, and organization in the literature. In particular, the general conception of space and phenomenological conception of place (see Chapter 6) due to the combination of ICT, space, and place have not been used to conceptualize the exploitation of geography. Similarly, software engineers' long-distance travels across sites to have face-to-face meetings is a very practical exploitation of geography, but has not been conceptualized adequately. This is because, again, its combination with earlier or later technology-mediated interactions for further relationship building has been under-theorized in existing publications. These instances show that the Newtonian rather than the Liebnizian view of space and place characterizes existing GSE coordination theory informed by the logic of indeterminacy. On the whole, geography is largely problematized, but its exploitation is lowly conceptualized.

Related to this problem is that materiality of technology is also weakly implicated in current conceptualizations that are informed by rational and indeterminate approaches. Various ICTs that enable various communication modes and information formats according to software engineers' various preferences and peculiar challenges are interpreted as instrumental rather than essential resources in existing publications. Such instrumental interpretations dominate the literature because ICTs used in GSE are esteemed largely as mechanical media that are used to manage uncertainties. The mechanical implications of ICTs are such that they are conceptualized in terms of linear information exchanges and reactions to various uncertainties in accordance with Claude Shannon's information theory. Their electrical implications – as in how they generate total field awareness as well as integration of software engineers, task and information through speed, synchronicity, and richness – are lowly conceptualized. Clearly missing in these publications is the cosmic consciousness among global software engineers; that is, their continuous experiences of closeness and interdependence, underscoring Marshall McLuhan's global village metaphor. The generation of cosmic consciousness is due to the material focus of ICT which must be interpreted as mostly electrical media. Until this interpretation is provided through an integration of the four perspectives (in this book), the limited instrumental interpretation of ICTs remains.

Having considered the thesis and antithesis to understand the low conceptualization of an integrated theory of coordination, we must proceed to a synthesis according to dialectical idealism. The best theory of GSE coordination must be an absolute idea that explains the practice whereby organizations combine organization, technology, information, and geography to coordinate software engineering activities. This idea of coordination must transcend but include the thesis of coordination by plans (informed by rationality), and the thesis of coordination by mutual

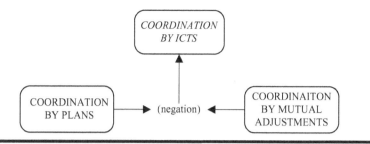

Figure 8.2   Dialectics of GSE coordination.

adjustments (informed by indeterminacy). And it must unearth new tactics that transcend standardization, modularization, variation, and socialization which are associated with rationality and indeterminacy approaches to coordination. The idea is coordination by ICTs (Figure 8.2).

Given that we have adopted a virtuality approach to propose this idea, we must first consider publications of coordination practice where coordination tactics informed by virtuality are referenced empirically but not conceptualized. Thus, concrete expressions of virtuality in GSE based on a combination of organization, technology, information, and geography are considered. The virtuality approach presumes the development of a technology-centered yet integrated explanation of coordination whereby the essential role of ICT in concrete expressions will be conceptualized. With the development of an improved theory of technological coordination in mind, the materiality of technology and exploitation of geography that have been overlooked in the existing conceptualizations will be conceptualized.

In the appraisal of the materiality of technology (Chapter 4) and exploitation of geography (Chapter 6), it was emphasized that materiality of ICT media is the structural basis of process virtualization. That is, ICT media has material structures that enable virtual integration of space and time as well as production and consumption of software. In this chapter, there are references to GSE publications showing the practical and instrumental roles of ICT media in standardization, modularization, variation, and socialization. But we now seek to show the essential roles of ICT media not as confined only within these classical tactics of standardization, modularization, variation, and socialization, but as the roles transcend these tactics. This pursuit should lead us to the conceptualization of synthesized technological tactics that are distinctive.

The virtuality approach to coordination in this book is premised on the understanding that GSE does not have only problem and fuzzy characters, but also has a distinctive discontinuous character (Chapter 3). How the problem and fuzzy characters are addressed has been adequately conceptualized through modularization, variation, and socialization in the indeterminacy approach (the antithesis). How the discontinuous character is addressed also needs to be conceptualized because modularization, variation, and socialization as coordination tactics are limited in the face of discontinuities. This limitation points to why technological interconnections are essential for GSE coordination. Indeed, the very thought and implementation of GSE would be impossible without technological interconnections. Without technological interconnections between developers at different sites, modularization, variation, and socialization are inadequate matches for this discontinuous task.

With the various ICT connections between developers at different sites, distributed software engineering team members concurrently achieve frequent, verbal, and informal communications in one site and avoid inhibited communication across sites.[43] Erran Carmel argues that stronger telecom infrastructure and better collaborative technologies are key forces for "keeping things together" in the face of spatial and temporal distances.[44] For instance, Lerina Aversano and colleagues report on a GENESIS platform (an integration of workflow management technologies with communication services) for coordination of distributed software engineering.[45] Even in situations where knowledge sharing, a coordination process, is used as in Julia Kotlarsky and Ilan Oshri's research,[46] they testify that effective ICT-based interconnections undergird this process for the achievement of successful collaboration among globally distributed software developers. Consider also the earlier references made in this chapter to the practical employment of various technologies, communication modes, and information formats by organizations to make coordination across discontinuities possible. All these empirical and concrete expressions point to the concept of connection as the coordination tactic most suitable for the discontinuous character of GSE.

How and why does connection as a coordination tactic transcend, but include, standardization, modularization, variation, and socialization to assume the status of a synthesized (primary and higher) coordination idea? First, connection translates global-distance into a virtual space of almost infinite opportunities for movement of information, tasks, and events.[47] It is the foundation for dynamic structuring that enables distributed software engineering teams to deal with non-routine problems such as changing customer requirements. Amid variations due to changing requirements, instant, simultaneous, and total awareness among team members is a critical requirement for coordinating dependencies in the software architecture. This cannot be achieved without ICT interconnections among them. For instance, the software architecture can be used to maintain standardized procedures more easily in non-virtual teams because of the availability of informal, verbal, and ad hoc communications among members.[48] But its standardized features are inadequate for easing the coordination process in GSE because task discontinuities are high and face-to-face communications across sites are minimal. Connection as a coordination tactic enables team members to have frequent, verbal, and informal interactions to maintain awareness across sites.

Second, connection is the synthesized coordination tactic for GSE because it is the carrier of technological information – information *as* reality[49] – which is the basic raw material for software production. The raw material of a task and its carrier define how its associated dependencies are coordinated. Because technological information is digital in nature, connection enables it to attain reach and richness at the same time.[50] Richness and reach of digital information through connection leverage all the classical coordination tactics (standardization, modularization, variation, and socialization) to manage site and cross-site dependencies at the same time. They create a strong representation or copresence of globally distributed tasks and developers. Richness and reach of digital information are tantamount to Marshall McLuhan's proposition that electrical media, unlike other media, enable total field awareness which is exemplified by simultaneity and synchronicity between software engineering actions and reactions. His ideas of global village and cosmic consciousness are clearly applicable to digital technology's role in GSE coordination. Therefore, standardization and variation are only adequate for coordinating this discontinuous task when they depend on an ICT-enabled connection.

If connection is the synthesized coordination tactic, then how and why can ICT be correspondingly understood as the absolute idea of GSE coordination? It is an absolute idea chiefly because of its essential digital principle. Jannis Kallinikos calls this principle "technology without matter."[51] This principle is primary for a technological explanation of coordination in this context because it gives existence and significance to ICT infrastructure. In particular, the principle of computer software is largely independent of physical and hardwired infrastructure. This independence separates its logical from its physical forms and distinguishes ICT from other technologies. The physical form remains inert until enlivened by the digital principle. The freedom of the digital principle enables it to transcend the boundaries and types of physical infrastructure to develop "toward the ether of the virtual and the immateriality of cognition and communication."[52] The ether of the virtual is a central aspect of the space of possibilities which is generated by applying the materiality of technology to the exploitation of geography. Moreover, the independence of the digital principle transcends the automation principle of mechanical media which manifests in linear and repetitive processes.

The digital principle of ICT is abstract but finds its concrete expression in the materiality of ICT, and then in how continuous quality improvement in global ICT infrastructure has been leveraged by organizations to interconnect people and create virtual software engineering teams. Therefore, problems of GSE coordination are traceable to problems of plans and mutual adjustments, but poor plans and mutual adjustments are not primary in this context because of the dominance of the digital principle of ICT both in the software engineering task (*qua* task) and for the task. The dominance of the principle is tantamount to its primacy as the cause of problems of GSE coordination; that is, coordination problems must be understood in terms of factors that lie beyond the physical constitution of ICTs. Coordination problems must be primarily understood in terms of this digital principle when the principle is not leveraged appropriately into efficient and effective connections among distributed software engineering team members. This understanding implies that solutions to coordination problems must also be sourced primarily from the digital principle of ICT. It is an understanding that should take our

Table 8.1 Coordination by ICT

Dialectics	*Organizational Logic*	*Coordination*	*Coordination Tactics*
Thesis	Rationality	By plans	Standardization
Antithesis	Indeterminacy	By mutual adjustments	Variation, socialization, and modularization
Synthesis	*Virtuality*	*By ICTs*	*Connection and capitalization*

thinking beyond the instrumental value of ICT to its material and essential values. Its instrumental value corresponds with its physical, mechanical, and rational character, while its material and essential values correspond with its digital character (see Table 8.1).

The side effect of variation is the prevalence of task conflicts, and it is easier to reduce task conflicts in collocated settings than to do so in global virtual ones. Therefore, in GSE settings, the co-existence of task conflicts and reduced environmental uncertainties is likely. The quest to address changing customer software requirements (a discontinuous task challenge) implies the accommodation of task conflicts that are engendered by variations in engineers' opinions on task procedures.[53] The one cannot exist without the other in the GSE setting which is characterized by discontinuity. However, variation is an inadequate concept for understanding discontinuous problems and how they are addressed because it pertains to understanding fuzzy problems. Therefore, there is the absence of a higher concept that is different from variation; such a concept is required for improved understanding of how discontinuous coordination problems are addressed practically. In GSE practice, Suprateek Sarker and Sundeep Sahay have reported a GSE team's actual experiences and practices in addressing discontinuous coordination problems:

> A number of strategies involving the use of technologies (e.g. the use of different media with varying degrees of social presence, and enabling the creation of a shared "placeless" space), organizational skills (e.g. creation and implementation of norms and improvisations, and appropriate distribution of work), and individual capabilities (e.g. cultural sensitivity and altercentricism, and the ability to adopt a polychronic approach) were discovered to be effective in dealing with the complexities.[54]

Such a practice points to capitalization of technology, information, and human resources – not just their variations. The required technological, information, and human resources that need to be capitalized for coordination are shown in the information management matrix (Chapter 5, Figure 5.2). The technology and human agencies for resolving tasks and bridging distance are discussed in Chapter 5. The relationships between technology and human agencies and their roles in resolving tasks and bridging distance point to the concept of capitalization being proposed here. There is high dependence on the dexterity of distributed software engineers to capitalize on technology and human agencies through exploitation of places and spaces in order to derive coordination benefits from distributed software engineering resources. Capitalization as a concept that transcends but includes variation, standardization, and modularization is the summary of coordination by using the materiality of technology for the management of information and for the exploitation of geography in the face of the paradox of GSE organization.

How and why does capitalization transcend, but also incorporates, standardization, modularization, and variation to assume the status of a synthesized (primary and higher) coordination concept? First, discontinuous software tasks often result in the emergence of unexpected challenges that do not have a match in the existing variation profile of a virtual team. When unexpected challenges emerge, they rather call for exploitation of an existing variation profile to address them. Capitalization as a coordination tactic is the process for preventing the transmutation of task conflicts into relationship conflicts which are detrimental for a team's performance of a non-routine task.[55] It is the fundamental reason behind the use of agile methods such as Extreme Programming and Scrum to achieve high morale among team members and strong customer commitment.[56] It also provides a better explanation for the success of GSE strategies such as follow-the-sun.[57]

Second, without capitalization, national and related socio-cultural diversities among virtual team members present problems of effective task performance.[58] The process of variation, based on coordination by mutual adjustment, is only capable of enabling a GSE team to react to discontinuous problems with these diversities. But the process of capitalization, based on coordination by ICTs, enables proactive handling of discontinuous problems. Variation is an adjustment process as compared with capitalization, which is rather a technological-intellectual process.[59] Capitalization reflects technology augmented intelligent behavior by distributed software engineers, but variation reflects their skills for adaptable behavior. This means that variation as a human behavior pertaining to the antithesis of coordination by mutual adjustments is limited without the inclusion of capitalization which belongs to the realm of coordination by ICTs. It also means that capitalization is the synthesized coordination concept.

Technology-augmented intelligence is the basis of capitalization because a GSE organization cannot achieve variations in resources without it. Hence, we can make similar arguments for why technology-augmented organizational intelligence is the foundation for the creative process for exploitation, as well as why it denotes a techno-intellectual explanation of coordination. Technology-augmented intelligence is needed to leverage software resources creatively and effectively.

Thus far, connection and capitalization are the two concepts that define the synthesized idea of coordination by ICTs (informed by virtuality). In this synthesis, the materiality of technology and exploitation of geography come to the fore along-side the paradox of organization and management of information. The emphasis of exploitation of geography is predominantly due to the materiality of technology. The very practice of GSE in the face of geographical, temporal, and socio-cultural distances is testament to actual coordination by ICTs that distributed software engineers use to exploit places and spaces. The strongest concrete expression of geography exploitation is in dataspace appropriation (Chapter 6, Figure 6.4). A dataspace is a digital information space that enables free data movement and access for mutual awareness and responsiveness to externally changing requirements. It is appropriated by distributed software engineers in the spirit of connection and capitalization to achieve coordination. There is a strong reflection of the general and Leibnizian conception of space in a dataspace. There is also a strong reference to McLuhan's total field awareness achieved by dataspace appropriation. General space conception and total field awareness pertain mostly to software product coordination across sites.

In the same spirit of connection and capitalization, global software engineers combine dataspace appropriation with workplace identification. A workplace is characterized by a combination of physical, innovation, and economic dimensions, as compared with place which has only a physical character. A workplace reflects the phenomenal conception of place because engineers identify workplaces with social meanings. A combination of dataspace appropriation with workplace identification is comprehensible only in the relationship between the materiality of technology and the exploitation of geography. This combination and corresponding relationship are weakly represented in extant theories of GSE coordination by plans and mutual adjustments. The reason for underrepresentation is that the rationality and indeterminacy approaches to explaining coordination have limited the conceptualization of geography exploitation and technology materiality.

Connection and capitalization tactics in the synthesized theory of coordination by ICTs do not conceptualize geography as only a problem to be overcome. Additionally, it conceptualizes geography as a resource to be exploited. Thus, while standardization, modularization, variation, and socialization are all discussed in the literature as tactics for addressing problems of geographical distribution of software engineering resources, connection and capitalization are discussed here as tactics (technology and human agencies) used to exploit geographical distribution. In the conceptualization of these tactics, technology is given an essential rather than an instrumental role because it is constitutive in GSE organization. It is constitutive because it is the only medium that enables the achievement of synchronicity, simultaneity, and representation that is critically required for teamwork GSE. This argument highlights the ontological difference between GSE and other globally distributed works. Software engineering needs a village ambiance to thrive because of essential requirements such as teamwork, community, and audio in order to achieve total field awareness and cosmic consciousness. Globalization of software engineering

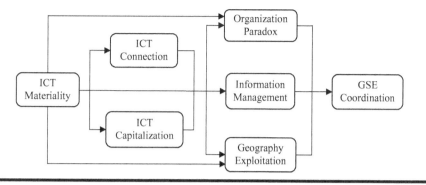

Figure 8.3   Technology coordination of GSE.

destroys these requirements. The sole restorer of these requirements is ICTs which generates synchronicity, simultaneity, and representation (see Figure 8.3).

In sum, a GSE organization is an epitomized global village constituted by the ICT medium which is electrical. In this book, the virtuality approach, corresponding to the proposed synthesis of coordination by ICT, points to the strength of geography exploitation and technology materiality as newly foregrounded perspectives (in addition to information management and organizational paradox perspectives) that provide new explanatory insights on GSE coordination.

Chapter 9

Illustration of Coordination

The proposed theory of coordination by information and communication technologies (ICTs) through the previous sections requires further illustration with an empirical example. The following empirical illustration shows how coordination by ICTs can be used to interpret process data from a longitudinal case study wherein coordination modes by a global software engineering (GSE) team were investigated. Process data from the case were deemed to be more appropriate for illustration than data from a survey or experiment because the data are more appropriate for answering the research question of how and why dialectics characterize coordination processes. The data are not being used to test the proposed theory of coordination by ICTs, but to illustrate its usefulness in explaining how the practice of GSE coordination reflects a dialectical process whereby ICTs constitute a synthesis of plans and mutual adjustments.

Moreover, the validity of this proposed theory does not depend on a representative proportion of illustrative data points for statistical generalizability across a large population.[1] Rather, it depends on adequate process data on necessary relations within social mechanisms[2] and real structures[3] that are used for analytical or theoretical generalization. Social mechanisms and real structures are exemplified by ICT, spatial distribution, and dialectics that are powerful and independent causal factors in this study. For example, global distribution of software development resources is an independent causal factor that is not essentially a feature of the empirical domain of the case. Therefore, although the necessary relations between these structures may be reflected in the case, those relations are found in other GSE contexts where these real structures and social mechanisms exist.[4] Therefore, these structures and mechanisms provide sufficient bases for generalizability of the proposed theory of GSE coordination.

Here is the background to the empirical case. Team Emrod (all names are represented by pseudonyms), a global software engineering subunit within a multinational ICT organization (SoftOrg) was upgrading a data mining application (also called Emrod) for remote data collection from its external customers' servers. This application contributed to the broader application (BigSoft) which was aimed at supporting SoftOrg's services to its customers. The support process was described by the team as follows:

> *A support strategy for the BigSoft Program is essential in the continuing efforts to provide quality service to our customers. This document is intended to define the support processes for the Remote Access Connectivity components within the BigSoft program. The BigSoft Technical Support Team responds to requests for help, information, or Service Requests, concerning the BigSoft product from support organizations in the Customer Support community in the EMEA,[5] Americas, Asia-Pacific and Japan regions. The Regional Response Centers function as the first line support organizations for BigSoft application-related issues. The Remote Support Technology Competency Center in EMEA, Remote Technology Support in the Americas and ... in Asia-Pacific provide second level support. For issues that cannot be resolved by the 2nd level super region support organizations, members of the 2nd level support teams can contact BigSoft Support for further assistance. Requests are sent to BigSoft Support by the submission of a support case using the BigSoft call submission web page ... These requests are assigned to a technical support engineer who takes responsibility for the call until its completion. The technical support engineer will work with the appropriate development lab engineer as necessary, in getting the issue resolved. The result could be a resolution, a workaround or a request for a change to be made to the application code.*

Several other subunits in SoftOrg (called release partners [RPs]) were involved in BigSoft development. SoftOrg hoped to achieve remote connectivity in which automated proactive data mining and diagnosing would occur in external customers' servers. It also hoped that it would achieve cost reduction by relying on SoftOrg's expertise around the world. Team Emrod was constituted of 12 members headed by a project manager (PM): three engineers and one architect based in Killarney, Ireland; one support person and one engineer based in Watertown, South Dakota; the technical lead (TL) and four engineers in Bloomington, South Dakota; and one product release manager based in Los Angeles, California. All 12 members reported to the PM who was also based in Killarney in the same work area with the other four. The time difference between Killarney and South Dakota is seven hours, signifying few overlapping hours of work between the two sites (Figure 9.1).

There were continuous fluctuations in customers' requirements that presented an unstable task environment to Emrod. Such fluctuations induced further changes in business requirements of RPs. The RPs were operating from sites in India, Brussels, other parts of the United States, and Britain. Thus, the team's work was discontinuous and was being affected negatively because these requirements served as inputs for Emrod development. Figure 9.2 below is a typical requirement.

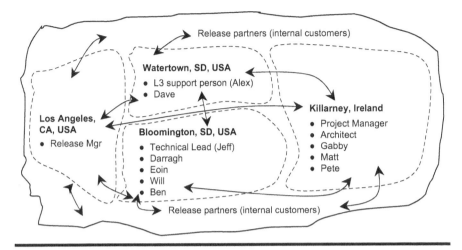

Figure 9.1   A sketch of the groups within Team Emrod and their locations.

Note: Arrows signify communications and all names are pseudonyms.

According to the PM, dependencies

> *between Emrod and release partners (RPs) [were] not that good; each part-*
> *ner [had] a different motive; commitment from them [was] not certain;*
> *engagement with them [was] continuous but the business requirements*
> *[could] be changed by a RP arbitrarily; there [was] competition for shared*
> *resources by RPs; dependencies [were] not smooth at all; business require-*
> *ments baselines are changing continuously in SoftOrg.*

A more significant fluctuation in Emrod's task environment was related to the highly critical nature of eleventh-hour changed requirements. In the early days of development, changing requirements were easier to deal with because there were enough time resources at developers' disposal. However, when the release was approaching, it was more difficult to deal with changing requirements because of the obvious time limitations. The following excerpt, which exemplifies an eleventh-hour requirements change, is the author's transcription of part of a teleconference that was being held three days before the first intermediate release.

The release manager (RM) and the PM become concerned about James, based in India, who had just sent an email to the RM with a set of new requirements.

RM: I'm surprised he [James] doesn't understand what the scope is. … Certainly, we've
got a lot of issues to be resolved and I don't know how we're going to resolve
that.
PM: I think we're good to go.
RM: We need to take them one at a time and get back to them.

BigSoft Requirement Submission Form			Submitter	Amos Lam
			Date	26 May, 20XX
			Business Sponsor	Onward Lam

One Line Description		Enable Support for Integrated Windows Authentication		
Detailed Description of Requirement	(from JAGae27922) If a customer is using a Microsoft Forefront GMT server, and requires that NT credentials are required before having access to the Internet, BigSoft (client & SPOP) will not be able to communicate with the SoftOrg Data Center. The NTLM authentication protocol authenticates the connection, not the HTTP transaction. Currently ISEE supports authenticating the HTTP transaction using either Anonymous/Basic or Digest Authentication – not NTLM/Integrated Windows Authentication. At the time of this submission there are numerous BigSoft customers unable to deploy the solution. Specific customers are listed below. The BigSoft team needs to provide a documented solution to customers with this network environment so that BigSoft is able to successfully be deployed.			

Partner Developed?	yes/no group:	Resource committed?	Platform Resource	Date of partner completion for submittal	N/A
	NO	If yes, BigSoft Platform Modification scope		None:	Medium:
				Low: X	High:

Specific Customers Impacted by Requirement	EDS, Chevron-Texaco, Fosters, Barclay's, PERN SA, numerous others

Impacts to program partners	Training: **Integrated Windows Authentication support reference needs be added to standard client/SPOP training modules**	Marketing: **None.**
	Learning Products: **The solution needs to be documented as a part of the standard BigSoft client/SPOP documentation.**	Other (specify): **None.**

Pilot Expectation	BigSoft coordination?	**Little/none. Limited deployment phase may be employed to validate the solution.**	Partner coordinated? (Name Program Manager):	No.

Business Priorities (specify applicable Priority for this Requirement): 1. Get the basics to work reliable and dependable first (fix what is broken, deliver the functionality of a minimum viable BigSoft) 2. Technology ready to massively deploy, pervasive deployment, remove obstacles (i.e. Predictive/HAO migration…, Customer Centered design.) support the deployment strategy. 3. Add functions/development that provide immediate visible additional revenues (i.e. contract compare) 4. Add functions/development that provide immediate visible cost savings (remote access, filtering, cost of maintenance) 5. Enable Global Delivery	This requirement applies to both priorities 1 & 2. The support for this type of authentication limits the total number of accounts BigSoft is able to deploy to and should be considered a base capability to support the Internet based connectivity.

Figure 9.2   Requirements request form.

PM: It's the same every time. I'm frustrated. We don't know upfront what we need to do.
RM: Everything gets up to the last week.
PM: We have the support agreement and so why is all this ...?
RM: It's gonna be an interesting one how we can answer these questions. I think we're almost guaranteed for rejection. We might get a conditional approval.

Because of the distance between them that reflected task and organizational discontinuity, the already erratic dependencies between Team Emrod and the RPs were worsened. For example, their output targets and evaluations were tied to the regulations of those sites, even though they were intermediate products that served as inputs to the development of Emrod. Team Emrod's inability to predict the changes in the state of business requirements was a typical instance of direct effects of the environment on its inputs.

These erratic inter-unit dependencies constituted uncertainties in the source of inputs for Emrod development because the engineers' coding had to align with RPs' own to facilitate smooth integration that would make BigSoft a success. Mutual knowledge problems were also typical uncertainties facing Team Emrod. For example, during my first meeting with the PM, he lamented about "guys making assumptions" in the early days of the project. Although the changing requirements close to release times were very critical and required high levels of developers' agility, the challenge also required high degrees of mutual understanding between them.

Illustration of Coordination by Plans

Against the backdrop of the task and organizational discontinuities, the team worked with the Emrod architecture (a high-level structural mechanism) as well as development methodologies, task schedules, release milestones, development platforms, meetings, and workspace arrangements (schedule mechanisms).

It was written in the *Team Handbook* that: "We meet once a week to discuss projects, status, outages, issues, etc. The meetings are usually on Monday or Tuesday, at a time appropriate for the time zones of the team members." The description of the BigSoft architecture was as follows:

> *BigSoft Remote Connectivity is the combined hardware and software that creates a secure VPN tunnel so SoftOrg Support Engineers can troubleshoot systems monitored with the BigSoft Advanced Configuration. At [the previous] release ..., the VPN tunnel began at the routers within the SoftOrg Data Center and ended at the VPN routers installed at the customer site. The Remote Connectivity architecture is shown below: [Figure 9.3].*

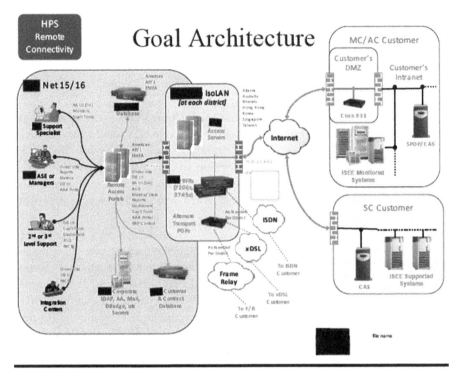

Figure 9.3   Architectural plan.

Moreover, the PM could exercise high-level control of these coordination mechanisms without supplanting developers' behavioral discretion at lower-levels. Thus, formal methods were used as a high-level structure that was

> *required because we interface with many external organizations that impose structure upon our team. However, within the team we use an iterative form of the 'agile' development methodology.* (Team Manual)

These mechanisms signify standardization in order to exercise control and ensure coordination by plans (including architectures and fixed arrangements of people and tasks in particular sites). These ensured large degrees of orderliness and regularity in the development of Emrod from the different locations. It was also strongly connected to the bureaucratic character of SoftOrg as a typical modern organization. Here is empirical evidence of linearity that exemplified coordination with mechanical media in order to manage dependencies, uncertainties, and conflicts pertaining to a problem task. But coordination by plans was not the dominant mode as compared with coordination by mutual adjustments and by ICTs.

Illustration of Coordination by Mutual Adjustments

Given the social and cultural diversity among Emrod team members and potential risk of mutual misunderstanding, the PM made arrangements for highly frequent interactions among them. For instance, the team members increased the number of their teleconferences and continued to engage in informal, verbal, and spontaneous interactions toward release deadlines. These interactions were important variation and socialization modes to coordinate a fuzzy software engineering task. Table 9.1 and 9.2 show notes on teleconferences held a few days before the second major release of Emrod.

These teleconferences were purported to provide Team Emrod engineers with total field awareness that was critically required for coordination of actions leading to the major software release. The updates provided by members during the meetings, as well as the knowledge and understanding that leads each engineer to take up a task to work on, constitute total field awareness. They also qualify as information management actions that made all the engineers conscious of the status of the

Table 9.1 Notes on a Team Meeting

Emrod Team Meeting Held on Wednesday 1st November 20XX (VSR in use)		
Time	*Notes*	*Remarks*
17:00	• *PM does a quick informal chat with Prince* • *PM asks Prince to write notes for him* • *Update on meeting Tracy's needs* • *PM: "We're good to go!" The next issue is training* • *Steve in charge of training – 2.00pm Killarney time, 7.00am Bloomington and Watertown time* • *Time demos assigned to one member; screen shots assigned to another* • *PM to Steve: "Are you going to record this?"* • *Steve: "I didn't get anything that shows what they want … Fortunately, we are always on top of our game"* • *PM: "We don't know what they want" – that's with a release partner called Singh* • *PM: "Any questions on the training … otherwise I think it's fine … we're are good to go with Bacon Phase 2"*	• *PM is leading the meeting* • *Tracy is one of the Release Partners*
17:38	• *Action list being discussed* • *PM goes through the list table and verifies that all is well*	
17:40	• *Round table led by PM* • *He requests that everyone should update himself in terms of BugZilla's bugs* • *250 bugs were unresolved yet (pushed to next phase)*	
17:53	*End of meeting*	

Table 9.2 Notes on a Team Meeting

Emrod Team Meeting Held on Tuesday 7th November 20XX (VSR in use)		
Time	*Notes*	*Remarks*
15:30	• *Some members complain that Jabber (instant messenger) is not working* • *PM is checking on who is available online* • *PM: "let me kick off"; after some informal chat with Ron* • *PM: "thanks to Prince for doing the notes last week. They're the best notes I've ever seen. I'll do it this time"* • *PM is sharing TeamPlan and going through the schedule and verifying to ensure that what is supposed to be done is done.* • *Each member explains (gives an update) on his work* • *PM: "All our lab expectations with respect to external release partners is done"*	*PM is leading the meeting*
15:45	• *PM opens project management page. He complains that it's not updated, and says "we're gonna get nailed by the auditors if we don't clean them up" by middle of the following week* • *PM allocates the updating tasks to members* • *Members chip in jokes about the PM not having upgraded his Internet Explorer to version 7* • *PM continues ... "The meeting page gives baseline information to the external release partners so that they don't engage me or Mike one-on-one"* • *PM talks about a "close-out perspective" on the upcoming release for external auditors* • *PM: "Release coming up at the weekend. We need telephone cover on Saturday morning"* • *Round table to see if there's something missing from what have been talked about* • *Frank is transferring to another team for 6 months. PM is arranging for a temporary replacement from Killarney*	
16:18	*PM announces my trip to Watertown and Bloomington in the United States to members*	*I went to Watertown and Bloomington about a week after this meeting to do interviews and observations at the two sites of SoftOrg.*
16:18	*End of meeting*	

entire Emrod engineering task and of every engineer's performance status. Because such cosmic consciousness and total field awareness are generated by the GSS, it qualifies as an electrical medium.

Besides the teleconferences, Team Emrod's continuous relationship building since the beginning of the development of Emrod had resulted in high mutual understanding. They exhibited high mutual understanding when they were dealing with eleventh-hour changing requirements. Only two Bloomington developers had met the Killarney developers face-to-face, so relationship building mainly within technology-mediated communications over time was the foundation for developing this mutual understanding.

Furthermore, traveling across the Atlantic, even by a few developers, had been very important both for sustaining high levels of understanding in cross-site inter-actions and for enhancing team cohesion and collective decision making. The engineers who traveled to meet face-to-face remarked that those encounters contributed significantly to their socialization. This was confirmed by one of the Bloomington-based developers:

> *I have [traveled to Killarney before], in fact. I've met all of them. It certainly does help to actually know what they look like because now you can put a face to the voice on the phone or a face to the email and say, "oh, that's Gabby." It makes it a little more personal knowing who's actually on the other end. Being able to see them adds something. Being able to actually sit across and talk to them whenever possible … it's good to get together with them so that you can talk about something other than work so that you know that you're dealing with another human over there. And that usually helps a great deal.*

Nevertheless, only two of the American engineers had met the Killarney developers face-to-face. Thus, it can be said that enhanced leadership by the PM constituted a strong foundation for socialization of the team members leading to mutual understanding between them. Leadership served as a significant compensation for the problematic effects of the numerous variations in Team Emrod's resource and behavioral profile.

Alongside the PM's leadership skills, there were several variations within Team Emrod's profile. First, the Bloomington developers had greater experience in remote connectivity application development and agile development than the others. Second, the seven-hour time difference between the American and Irish sites was also symbolic of temporal variation. Third, there was a difference in Team Emrod's behavioral profile as evidenced by the Team philosophy and methodology:

> *fast, lightweight, nimble … Do the Right Thing … at the expense of "the process" … Our engineering methodology is a combination of a larger, traditional phased approach for use in outward-facing communications,*

> *and an internal iterative "agile" methodology for use within the team. The larger methodology is required because we interface with many external organizations that impose this structure upon our team. However, within the team we use an iterative form of the "agile" development methodology.*

Fourth, socio-cultural variation was witnessed in the manner of their interactions. For instance, a Killarney developer reported that "we are more social in Killarney," while the "Americans hone in straight into the task." He said that the Irish infer a lot from conversations, while the Americans are kind of "black-and-white" in their talking. According to the PM, these cultural differences resulted in "guys making assumptions."

Concerning the use of places and spaces, the engineering of Emrod was modularized physically into location-based components that were integrated eventually to result in the final product. The TL, who directed team affairs in terms of technical issues, designed the tasks to be less interdependent across the three sites. He said:

> *one of the things I tried to do in terms of task interdependencies as TL is to minimize those interdependencies especially between Kerry and Bloomington and Watertown ... I tried to design the tasks so that they are completely independent between the regions. I would not necessarily actually do that if it's between two engineers on the same site.*

According to him, the tasks had to be independent "otherwise they would take very very long times to finish." His decision was not due particularly to the seven-hour time difference between South Dakota and Kerry, but to spatial differences between the three main locations. In addition to this explanation, I learned that the Bloomington site engineers worked on the connectivity and data mining component of Emrod while the Kerry site worked on its security component. I also observed that, at each of the three sites, the engineers occupied cubicles that were grouped within a 400-meter square area. The TL also said that even though Watertown and Bloomington had all their working hours overlapping because they were in the same time zone, "we still had to separate the tasks just because we're not physically together."

Illustration of Coordination by ICTs

The collocation of task components and engineers in order to manage dependencies (as part of coordination by plans above) caused a problem of low dependencies between the sites. Low dependencies led to mutual knowledge problems across the entire virtual team and difficulty in working together toward the final product.

For instance, the TL was asked about the effects of collocating task components on cross-site dependencies. His response was that he was aware of the communication and mutual knowledge problems across sites; and described how they addressed them in the following words:

"Now when it comes to decisions about how to do [the tasks], we do collaborate – we try to force the people to come together on them."

The team was forced to come together through numerous cross-site interactions using various technological media. Apart from the weekly teleconferences involving all engineers, there were several other interactions in the form of phone calls, one-to-one and broadcast e-mailing, one-to-one and many-to-many instant messaging, posting of documents on the document management system, and assignment of bugs that was done by posting the bug details on the bug tracking system. One support person, who was based in Watertown and was the main link between the users (of Emrod's software) and the team, also relayed user-reported problems either to the appropriate subgroup or to the entire team to get them resolved. Such resolution, both at the subgroup and team levels, engendered several cross-site interactions to enhance mutual support between collective decision making and team cohesion.

Collective decision making focused on dealing with both internal and external facets of the engineers' work. Internally, the reduced task interdependencies threatened to reduce mutual task awareness. High levels of mutual task awareness were important because they ensured that operational problems were resolved early so that they did not become more difficult to resolve during eventual integration. Besides, high degrees of collective decision making were important because, periodically, the engineers faced strategic challenges that called for guidance from other engineers (especially the more experienced ones in the United States) to achieve resolution. Externally, continuously changing requirements from release partners and customers threatened to cause instability in operations. Such instability required the team's collective responsiveness through continuous team-level interactions.

Team cohesion reflected the need for relationship development among Emrod's engineers, and it could be achieved only through continuous interactions aimed at task awareness and collective decision making. Because of the reduced task interdependencies, the TL's design suggests that there was potential danger for team members to exhibit inadequate mutual knowledge about their perceptual differences. Adequate mutual knowledge was crucial for increasing mutual understanding of information exchanged between them. As indicated in the reported instance below, the team drew upon this mutual understanding to remain responsive to changing requirements.

We have a series of what we call actions that our software can perform. And we have authorizations as to who can perform those actions. Right now we have a spreadsheet that says what all of our actions are and what they do. But that's too complicated and we need to simplify that, and we've

> *got multiple names for the same kind of actions. And it turns out that these two names are really meaning the same thing and we need to bring them together and so the complex problem is "what do we want to call these and how do we want to coordinate our changes to our code in our database to resolve this issue?"*
>
> *This is going to require a collective phone meeting. Some of these problems you look at them and you might decide, I want to recommend a particular solution. And you come up with all the alternatives and you write it up and send it off in an email to the rest of the team or the team members who are interested in this particular area. And then they'll reply back with what they think, and you can narrow it down that way as well over email, but that usually takes longer. That takes days as opposed to an hour or so.*

This is a typical instance of dataspace appropriation through the combination of ICT and space, reflecting the general conception of space. Here, exploitation of geography and materiality of technology are manifested.

On the whole, the task and organizational discontinuities, low dependencies, and high uncertainties between the sites were primarily overcome by the diverse technological connections between the sites. They could not be overcome only or primarily by the standardization mechanisms, but were overcome by the repertoire of telecommunication media that was made available to the team by SoftOrg: a teleconference system, e-mail, instant messenger, telephone, and a software bug management platform called Bugzilla. Members of Team Emrod worked with these media extensively to have regular interactions among themselves. But coordination by ICT, through connection and capitalization, was concretely expressed in how the GSS (which embodied all these media) was used to manage cross-site dependencies, emergent and varied uncertainties, and task conflicts.

The teleconference technology connected all Team Emrod members through phone lines complemented by a virtual meeting room, remote desktop sharing, and instant messaging software. Headsets-with-microphones were used in voice conversations instead of the normal phone handset. A developer joined a meeting by dialing a common number to an automated instructor which would instruct the dialer to enter a common conference code. This established a voice connection with all other conference participants. Then, using secure conference keys received from the conference presenter either beforehand or immediately after establishing voice connection, the developer executed the software that opened up the meeting room and availed of instant messaging and remote desktop sharing to the presenter and participants. At this point, a connection for voice, text, and images exchange among participants was established, and the meeting commenced. The technology did not transmit video signals.

The interface of MeetRoom displayed textual representations (and not pictures) of all people who were participating and distinguished clearly between presenter(s) and participants. As Figure 9.4 shows, the interface showed all people online, it

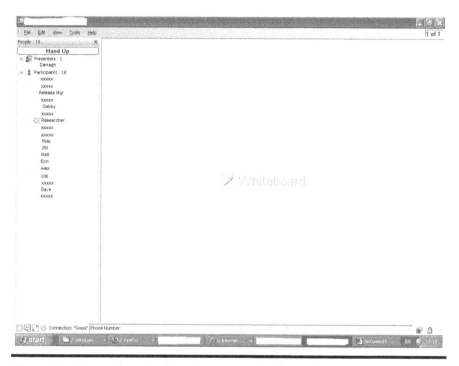

Figure 9.4   Screenshot of teleconference application.

distinguished clearly between presenters and participants, and it identified the person who was sharing a document.

A being-shared document (and hence, interface) would open in a new window; and the sharer would have the sole authority to modify any documents being shared. Modification of being-shared documents consisted of highlighting, deleting, and inserting particular texts or images according to suggestions from participants. Switching of document sharing was agreed in verbal communication to the hearing of all attendees (Figure 9.5 and Figure 9.6).

Table 9.3 shows notes on a software testing teleconference held one month before a major release of a version of Emrod, showing a record of coordination by the distributed engineering team.

The team's ability to interact and gain awareness across sites was achieved by their use of these media. They were applied at various times to match parameters such as the detail of information needed; the reckoned length of the communication; whether the communicator wanted the communication to be obtrusive or unobtrusive; the necessary number of people who needed to get the information being communicated; whether the information needed to be stored or not; and whether the communicated issue required an immediate or delayed response. The following notes taken during a teleconference, which at that time became the

```
I sent the following mail to Ken and the PMMS/Libra
team:

-----Original Message-----
From: Green, Aaron
Sent: Wednesday, June 21, 2006 3:09 PM
To: Wright, Ken
Cc:
Subject: RE: Scorecard Metrics data feeds


Ken,

I've gotten the code ready to test from one of our test
servers. The name of the system is ....net and the ip
address is 16.112.139.255.

That system is called a RCTS or Remote Connectivity
Toolbox Server. You can connect to the system using a
web browser, such as Internet Explorer or Mozilla
Firefox using https:

                    https://....net

You will need either a Class A or B Digital Badge
loaded into the browser to connect to the RCTS.

After you get logged in, go to the "Reports" page which
is on the navigation bar at the top of the page.

On the "Reports, Notifications, Data Exports and
Subscriptions" main page, you will see reports that can
be viewed, as well as notifications and then data
exports.

If you click on one of the data exports listed, it will
generate the information and display it on the screen.
Scroll down to the bottom of the screen and click on
the "Mail Now" button, which will then email this data
export to your email address.

Do this for each of the data exports listed and verify
that the format of the data is correct. James gave me
the format of the four data feeds that he is providing
to you and I have re-created those formats.

Since this system is a test system, some of the data
may not be as complete as it would be from a production
system.

When the ... project is put into production, which is
scheduled for Saturday September 9th, we can have L3
Support subscribe GDIC_PMMS_Bangalore to all 4 of the
data exports, or individual team members can go to the
production RCTS and subscribe themselves.

Let me know if you have any problems or questions.

-Aaron
```

Figure 9.5 Sample e-mail on data exporting testing.

I received the following from the PMMS/Libra team
regarding the testing of the data exports:

-----Original Message-----
From: C D, Ian Kelly
Sent: Thursday, June 29, 2006 12:22 AM
To: Green, Aaron
Cc: GDIC_PMMS_Bangalore
Subject: RE: Scorecard Metrics data feeds

Hello Aaron ,

The findings of the Scorecard links are as follows
• The link is working fine and all the Data Extract
      hyperlinks are active
• For metric I133 the Hyperlink is Active RAP Users
      Per Country
• For metric I143 the Hyperlink is Companies
      Registered on RAP Per Country
• For metric I134a we need two files and the links
      are Unique RAP

Customer Connections Per Country and RAP Customer
Connections Per Country. But the ordering of the data
fields are changed, so kindly change them as given
below:

Data source, Date, Entity, PL, Unique Customer
Connections - for Unique RAP Customer Connections Per
Country (This is given Incorrectly in the Webpage)
Data source, Date, Entity, PL, Outbound Connections -
for RAP Customer Connections Per Country (This is
given correctly in the Webpage)

We also wanted to know if it is possible to change
the hyperlink name as follows, so that we can
recognize the difference in the names of the files:
Unique RAP Customer Connections Per Country: unique
RAP Customer Connections Per Country Split
RAP Customer Connections Per Country: Outbound RAP
Customer Connections Per Country

Hence kindly make the corresponding changes and
inform us so that we can check and confirm it to you.

Awaiting your early response in this regard.

Cheers !!

Ian

Here was my response to Ian and the team:

-----Original Message-----

From: Green, Aaron
Sent: Thursday, June 29, 2006 10:50 AM
To: C D, Ian Kelly
Cc:
Subject: RE: Scorecard Metrics data feeds

Ian and team,

Thank you for testing the data exports.

Yes, I will make the changes you requested, that is
not a problem.
I will be off tomorrow June 30th, and Monday July 3rd
and Tuesday July 4th are holidays in the US. I will
work on the changes when I am back in the office on
Wednesday July 5th and inform you when they have been
completed so you can verify the changes.

-Aaron

Figure 9.6   Sample e-mail exchanges on data exporting testing.

Table 9.3   Notes on a Software Testing Meeting

BigSoft A.04.75 Phase I CloseOut Meeting Held on Thursday 17th August 2006		
Time	*Notes*	*Remarks*
15:05	Aaron asks if Frank has feedback forms on what tests and what experiences have occurred	Feedback: what tests? what experiences?
15:08	Aaron begins to share screen	In all bug (software defect) instances, Aaron read the whole bug description even though he had displayed or shared it with all on the screen
	Bob talks about his test for which Rick had earlier sent a report on Bugzilla	
	Bob shares the report on the screen	
	A lady agrees with Bob	
	Aaron rephrases a question. A clarifying question because others did not seem to understand	
15:14	PM asks about the necessity for a single control unit	
	PM: "Default if participant only"	
	Hunt: "what if we change the participant to unchecked"	
	Discussions are all about "default and control." Everyone contributing	
15:17	New shared screen being discussed. It's about RAPS special instructions	RAPS – remote access portal server
15:24	Discussions about connected customers' system	
15:25	Discussions about MS – Telnet F-Secure test	
15:26	Someone is testing from a laptop	
15:27	Chat facility (ad-hoc remote access) being discussed …	
15:28	Alicia explains Digital Badge login … Ian comments … The issue takes a longer time to settle	
15:35	Remote access still being discussed. Focus is on customer system information	
15:37	Scenario 3.1.1 (a figure) being shared	

(Continued)

Table 9.3 (Continued)   Notes on a Software Testing Meeting

BigSoft A.04.75 Phase I CloseOut Meeting Held on Thursday 17th August 2006		
Time	*Notes*	*Remarks*
15:39	*RAPS special instructions being discussed again*	
15:43	*Discussion of ad-hoc remote access chat facility continues*	
15:44	*Discussion of installation of ad-hoc remote access client software on Windows Server*	
15:48	*Sharing participants' desktop*	
15:55	*Discussion about multiple contact e-mailing in Emrod …* *Aaron: Data Exports not yet tested for PMMS/ Libra team*	*On PMMS/Libra team, see Figure 9.5 for a sample e-mail sent by Aaron to Ken and the team in India concerning a software bug (Bug 1893) that has to be tested. Aaron had sent the mail one month before this meeting.*
	• Someone explains to Aaron that some acronyms did not convey any meanings to users *• Aaron responds that he could have got them wrong.* *• Someone says if we can explain time stamp means connected, connections means connecting (for example), then we can be clear*	*Acronyms - headings*
	Ian Kelly (an Indian located in India) clarifies his point to a request by Aaron to respond to the testing of the Data Exports *No problem with understanding Ian's message delivered with an Indian accent*	*See Figure 9.6 for an earlier e-mail exchanges between Aaron and Ian on Data Exports Testing which are being referenced in this meeting.*
	Meeting has so far been very formal and exclusively devoted to the test results	
16:17	*PM thanks all participants for contributing to the test*	
16:18	*End of meeting*	

primary means for managing software bugs ahead of standardization and variation, illustrates this:

The PM talks about the "bugs so far"
Bug 2489 is particularly problematic.
On another bug, a member clarifies a situation
Bug 2873 which says "the screens are difficult to navigate" is discussed
The PM asks whether they have to push back 2873
The others do not agree: one says the problem is about page size enlargement for visually
*   impaired against page clutter*
PM: Members P and Q, "will you work on that and get back to me?"
Bug 2820: A member re-arranges back and forth between its severity and status to
*   explain his point – they all count how many are severe against the rest*
"That leaves with 32 other bugs, 20 of them open"
PM: "Do you wanna stop sharing for a sec?" (he wants to share something else)

These notes demonstrate the team's combination of telecommunications media to work with a connection mode across sites to coordinate work. It also demonstrates the primacy of ICT for coordinating the bug problem resolution because the absence of ICT would render the plans and mutual adjustments ineffective. The repertoire of telecommunication media ensured the presence and utility of the connection mechanism that managed the problem of low dependencies and high uncertainties across sites.

Concerning capitalization, the variations by themselves would have undermined the team's coordination because they could cause task and relationship conflicts among members. However, the variations were further capitalized by the team to deal with the complex character of the development of Emrod. The following instances testify to how it intelligently capitalized software resources to achieve coordination.

When customers brought new requirements at close-to-release times, team members used the different technologies to call upon colleagues who had higher levels of development knowledge and experience. The communication mode depended on the nature of the problem and the explicitness of the information required. Typically, developers used instant messaging for very short queries, they used telephone calling for queries which required more time for interactions, and they used e-mailing when the explicitness of the expert's response demanded a corresponding explicit query. Table 9.4 presents notes on a readiness meeting for an impending major release of that version of Emrod referenced in Table 9.3.

These are instances of the team's intelligent capitalization of globally dispersed resources and ICT for task coordination. This intelligent act was primary for coordination because without it, the plans and mutual adjustments (through standardization and variation) would be ineffective.

Table 9.4   Notes on a Software Version (Bacon) Release Readiness Meeting

Emrod Bacon Readiness Meeting Held on Monday 21st August 2006 Led by Frank Johnson		
Time	*Notes*	*Remarks*
16:03	• *Frank asking if today is a bank holiday in Europe* • *Not sure about how many have logged on; he asks "who do we have on the phone?"* • *Jimmy arrives and confirms identity and clarifies that bank holiday is next week* • *Frank says the meeting is "just to keep track of things"* • *Frank: "there is a tone when someone joins the VR." The tone notifies existing members of the arrival of a new member* • *Through the VR (virtual room) as well as verbal queries, Frank, the moderator, ensures that members are present before commencing meeting*	*Meeting is expected to be short*
	Meeting begins with discussions of "deliverables" *Bacon calendar being shared by Mike* *Frank confirms bank holidays in Ireland and UK coming up and incorporates them into plans*	
16:18	*Frank: "It will be nice to give somebody a phone call." … "We don't have a progress on something which is crucial for release count next week." … "We don't have a foot to stand on"* *Someone requests that they contact Betty Zhang on an important component on the release. The person had to spell her name before Frank got it.*	
16:22	*Frank shares Bacon TSG Action Items testing, and discusses them to clarify those items which are completed/approved or not*	*TSG – Technical Support Group*
16:25	*Frank shares new screen – an e-mail message sent by Brian Cox with an attachment on VSR Lab and Field Testing* *Brian discusses it to clarify aspects of it*	*VSR – virtual support room* *The test had been done on 7th August (two weeks before this meeting)*
16:27	*PM jokes about Shaun's athletic abilities and all or most of them laugh for about 5 seconds, and then they get back to business*	

(Continued)

Table 9.4 (Continued)   Notes on a Software Version (Bacon) Release Readiness Meeting

Emrod Bacon Readiness Meeting Held on Monday 21st August 2006 Led by Frank Johnson		
Time	*Notes*	*Remarks*
16:29	Someone has notified Frank that he's stopped sharing the page after he had done so for about 30 seconds Frank explains that "it dropped off" PM talks briefly about RPs, but postpones detailed discussions to "tomorrow evening" PM says it is important to "remove dependency" Something has happened that has removed an item from members' calendars Frank will ask Tracy to get clear requirements from "the Shaun guy"	Shaun, a remote partner (RP) is the source of dependency and his questions after questions after questions removed
16:37	Frank shares a new page, then stops sharing it after 20 seconds, and then resumes sharing after 30 seconds	
16:40	Frank: "Anything else"	
16:43	"To do pilot" PM: "what does that mean"	
16:46	Aaron shares UAT summary ... Participants continue talking about desktop sharing in Bacon	UAT – user acceptance test
	PM asks is Prince Hunt is online PM: "may be we follow up" Prince: "I can prioritize the UAT ad hoc ..." PM: "any dissenting voice on that"	
16:57	PM: "are we supporting Windows 2000?" Prince: "I think W2K is still widely deployed" PM: "Aaron, do you want to pull up everything we've talked about on UAT?" Aaron is trying to share Bug list with participants, but PM says he's "got it" – but Aaron shares it anyway PM is conscious of the UAT list because he wants to make sure he doesn't miss anything. He's going to check that with Aaron later	Later – in another meeting
17:05	PM: "what's the next meeting?"	
17:06	End of meeting	

Uncertainties generated by cross-site dependencies required the use of different technology-mediated interactions between the sites. This is how a developer explained the team's response to a customer's last-minute requirements:

> *In that case, it was mostly emails. Jeff starts an outline. Ok, this is what we need to do, and here's what everybody's assigned to do, so go off and do it. And then we would send an update to the whole list; or you just reply all and say ok I've got my part done and here it is. If we had a team meeting scheduled between [the time we learn about the changed requirement and release day], then we would discuss it in our team meeting. We usually didn't have a scheduled phone conference between the team, just the e-mail—broadcast email [using the team mailing list].*

The time difference was intelligently capitalized by the team to manage uncertainties, especially those that required more collective and sequential actions across sites to address. Members therefore took advantage of the difference to undertake sequential actions. For example, Killarney developers would work on aspects of the problem while South Dakota developers would be sleeping; then when Killarney developers close from work, South Dakota developers would take over actions on the problem; and then Killarney developers would go to sleep and return to the problem the next day, and so on until the problem was resolved. Furthermore, they exercised discretion to adopt agile methods for development so that they could deal with their customers' changing requirements. Although SoftOrg's regulations generally required the adoption of formal methods, the team's capacity for agile development within operational cost limits was crucial for dealing with such uncertainties.

In sum, the connection and capitalization modes incorporated yet transcended standardization and variation in the coordination of Team Emrod's discontinuous task. The combined use of developers' experience, use of ICT media, agile methods, and capitalization of differences to follow-the-sun all point to the team's coordination of resources with techno-intelligence. The connection mode was enabled by the computational logic undergirding various communication media and pointing to the fact that ICT is the absolute idea of GSE coordination. We also saw that coordination by ICTs was a creative act by Team Emrod members as they combined and re-combined digital logic to organize the various ICT media and to coordinate their discontinuous task. All the technology choices made by Team Emrod members, the functions of the technologies, and the diverse communication genres were used to draw upon various combinations of digital logic. The team's coordination by ICTs certainly included plans and mutual adjustments, but ICT was primary because it incorporated and transcended plans and mutual adjustments.

Chapter 10

Reflections

This book began with the premise that existing global software engineering (GSE) coordination theories are characterized and limited primarily by rationality and indeterminacy logics which are not only traditional but contradictory. To address the limitation, a different logic of virtuality has been used as a new approach to the development of coordination theory. The virtuality approach is justified by the predominance of information and communication technologies (ICTs) in GSE, where they play practical roles as the task itself, as structure, and as function. Based on this approach, GSE coordination has been interpreted as a dialectical process in coordination theory. The interpretation shows that coordination by ICTs negates coordination by mutual adjustments and plans. Thus, the absolute idea of coordination is ICTs has been argued in this book to transcend yet include mutual adjustments and plans. In drawing our attention to technology coordination, this book argues that previous explanations esteem ICT only as an instrumental rather than theoretical issue. This book shows that ICT is an epistemological issue in GSE coordination because it explains how and why ICT resources that enable virtual organization are also essential for coordination. On the whole, the focus on technology coordination with a virtuality approach has contributed a new explanation to make us rethink, research, and redo GSE coordination.

This book's claim for theoretical novelty is due to the deeper and broader explanation of coordination practice (as compared with previous ones) being proposed. The argument for greater depth in this explanation of GSE coordination than previously is first based on the virtuality approach. Virtuality is the core logic underpinning the relationships between technology, information, geography, and organization. In previous explanations of coordination where virtuality has not been the approach, no essential technological explanation of coordination has been developed. In essential technological explanations, behavior is ascribed to the materiality of ICTs rather than their instrumentality. But previous literature is dominated by instrumental explanations which do not trace coordination practice to the logic of virtuality. Because these instrumental explanations are not approached with virtuality, they lack essential technological arguments. Yet the reality and practice of GSE coordination across the world, as testified abundantly in the literature, are such that ICTs are not just supportive but essentially generative and transformative resources. The GSE work configuration itself is evidence of how ICTs generate and transform organization. Its generation and transformation powers have been one of the most interesting research subjects in information systems and organizational studies since the last quarter of the twentieth century. Writing in their edited volume, *Materiality and Organizing*, Paul M. Leonardi and colleagues point out that, "new technologies bring changes to the way people communicate, act, and organize their social relations."[1]

Thus, the espousal of a new explanation of GSE coordination by ICTs with a virtual approach is in harmony with what the literature has long espoused before: that computer-based collaborative tools are vital for GSE.[2] But beyond the harmony, this approach traces coordination to its core technological foundations where technology is the primary and synthesized idea rather than an epiphenomenal supportive instrument. In using virtuality to take this deep step to proffer this explanation of technology coordination, this book's contribution is similar to Ann Majchrzak and colleagues' explanation of technology adaptation to team structures,[3] and with Arvind Malhotra and Ann Majchrzak's theory of technology adaptation to knowledge sharing.[4] Interestingly, these are the only known publications of technological explanations in virtual teamwork. However, adaptation is practically and theoretically different from coordination. Adaptation is about alignments and misalignments between technology, delivery systems, and performance criteria,[5] while coordination is about managing dependencies.[6] Technology coordination of GSE also extends beyond adaptation because of how it has been interpreted in this book as a dialectical process that includes adaptation.

A technological explanation of coordination (or an explanation of technology coordination) is an important theoretical contribution to the body of knowledge of GSE because it adds a new logical foundation to our understanding of information systems, which is a combination of ICT and organization. Based on the media ecological analysis of how the materiality of the repertoire of GSE technologies enables coordination, we can understand why certain virtual indices of coordination are attributable to technological generation. The virtual indices are technology connection and capitalization as coordination concepts – concepts that are visible expressions of coordination by ICTs (being proposed here as the absolute idea of GSE coordination). These virtual indices show how the materiality of ICT combines with places and spaces to enable the conversion of geographical distance, which has previously been theoretically esteemed as a hindrance, into an opportunity. These indices provide knowledge and understanding of how and why ICTs generate village dependencies and dataspaces which in turn enable software engineers' creativity.

The knowledge of GSE dependencies as village dependencies generated by ICTs among global software engineers is a distinctive proposition in this book. Village dependencies are technologically generated and managed. They are replicates of collocated software engineering because the virtualization that leads to GSE is intended to create a simulated collocated environment. Knowledge of village dependencies in this book traces their origins to ICTs, and then explains how ICTs dominate in their management. James D. Herbsleb and Rebecca E. Grinter's research on *Splitting the Organization and Integrating the Code*,[7] which traces dependencies to product structure, is an implicit explanation of the technological origins of dependencies. Drawing upon Conway's law, they trace the organization structure of GSE to the product structure. Although the product in GSE is technology, their argument in the first place is not explicit. Second, their argument is limited to the technology product at the expense of other technologies such as GSS that generate the total field awareness and cosmic consciousness required for successful coordination. Other previous theories of GSE coordination overlook the technological origins of dependencies in spite of extensive references to dependencies in the literature. Even the research by James Herbsleb and colleagues on *Distance, Dependencies and Delay in Global Collaboration*[8] is also devoid of technology generation of dependencies and enablement of their management. In their research, they explain the degree of dependencies at different GSE sites, which is laudable, but their ideas on dependencies and their management preclude technological origins.

Beyond the knowledge of technologically generated village dependencies, we have also learned afresh about technology-generated management of dependencies in the proposed theory of coordination by ICTs. Managing dependencies in GSE with architectures and fixed structures (coordination by plans) is helpful as far as our knowledge about mechanical aspects of the work configuration is concerned. Similarly, managing dependencies with software engineers' autonomous behavior, long-distance travel, and face-to-face meetings (coordination by mutual adjustments) is also helpful as far as our knowledge about human aspects is concerned. The espousal of managing dependencies with simultaneous, circular, and synchronous connections through the exploitation of space and ICT provides new knowledge about its electrical and digital foundations. The mechanical and human aspects of dependencies management characterize previous theories of GSE coordination (see Chapter 5). Of course, where the technological origins of dependencies are overlooked in coordination theory development, where the concept of dependencies is not conceptualized in terms of materiality of ICT media, we should not expect knowledge of managing dependencies to be electrical and digital. Instead, we should expect the role of technology to be conceptualized as instrumental or supportive. In this book, the combination of technology and geography as the basis of managing dependencies has been conceptualized by highlighting the electrical and digital foundations.

The knowledge of dataspaces generated by the combination of space and ICTs is also distinct because it is absent in both name and conceptualization in previous theories of GSE coordination. Knowledge of dataspace utilization also signifies greater depth than previous conceptualizations of geography where the role of geography has previously been limited to places at the expense of space. For this reason, the geography of GSE coordination has been explained as a hindrance rather than an opportunity. But space is always an opportunity for all endeavors including transportation, construction, and organization. The development of GSE work configurations testifies to the practical exploitation of space, but this has not been conceptualized in previous research. In this book, the virtual approach, including media ecological and innovative geographical analysis, has revealed the concept of dataspace which is generated by the combination of ICT and space. The dataspace is a technology application layer of GSE coordination that is separable analytically from its technology infrastructure layer. The technology application layer reflects the general conception of space which is largely unhindered by physical constraints. Epistemologically, dataspace reflects an aerial view of GSE coordination which complements the extant limited terrestrial view expressed in physical geographical terms.

The virtuality approach has also facilitated integration of the previous perspectives on coordination (information, technology, organization, and geography). The integration which is concentrated in Chapter 8 is the main premise of this book's argumentation for greater breadth in explanation (than previous ones) of GSE coordination practice.

One of the arguments for breadth is that the virtuality approach adds a new fundamental logic to the explanation of coordination. Previously, virtuality has been kept at the supportive level of explanation rather than at the fundamental level, and so it has hardly been esteemed as comparable with rationality or indeterminacy logics. The supportive estimation of virtuality in previous explanations has produced knowledge of its actuality or manifestation. Thus, the idea that GSE teams are virtual and generated by digital technologies is commonplace in previous publications. But the idea that virtuality is itself the generative or real mechanism that gives rise to GSE teams and to coordination of their activities is not clearly articulated in the literature. In terms of Aristotle's theory of causality,[9] virtuality has been argued in this book to be the final cause of GSE coordination. The final cause is the telos, the teleology, "that for the sake of which the thing is done."[10] Therefore, rather than thinking about global software engineers coordination practices primarily as efficient action (rationality) or effective action (indeterminacy), this book proposes that we should think about their practices as primarily creative action (virtuality) (see Chapter 3). Virtuality expressed in creative action is the primary teleology of GSE coordination – coordination because of virtuality. Consider how digital technologies "without matter" combine with space to provide infinite opportunities for action, and then see how that underscores creative action in the virtuality teleology. In sum, GSE coordination is to be understood primarily as the pursuit of the final state of virtuality, in addition to the previous secondary understanding of coordination as the pursuit of rationality and the management of indeterminacy.

Another argument for breadth is that this book proposes a more comprehensive knowledge of the relationships between the four perspectives. Concerning the breadth of technological relationships with the other perspectives, it has been shown in the previous chapters that GSE coordination is achieved because the materiality of technology is used for management of information and for the exploitation of geography in the face of the paradox of GSE organization. Technology is the material basis of coordination as it enables the exploitation of geography; technology agency combines with human agency for management of information, and technology is essential for the spatial and temporal resolution of the paradox of organization. Among the other three perspectives, the paradox of organization facilitates information management because the loose coupling between GSE resources and activities provides the flexibility for the enactment of various combinations of technology and human agencies. The exploitation of geography, witnessed in the pursuit of GSE itself plus dataspace utilization and appropriation, upholds organizational paradox. Exploitation of geography leads to loose coupling between dataspaces which reflect the general conception of space and workplaces that also reflect the phenomenological conception of place. The application of these conceptions to coordination shows that exploitation of geography leads to information generation, processing and sharing at particular sites and across sites.

These relationships between the four perspectives, and how and why they constitute a more comprehensive explanation of coordination, are a progressive step beyond the particular explanations in previous literature. Previous literature is characterized by fragmented explanations which have been integrated by this book. The integrated explanation is a significant contribution toward the science of GSE coordination which is an aspect of administrative science. James D. Thompson, for instance, wrote that for administrative science to advance and be distinguished from administrative lore, researchers must focus on "relationships among phenomena under stated conditions"; we must "simplify understanding of relationships through use of abstract concepts which permit generalization"; and define concepts "by a series of operations which permit a sensory perception and identification of the phenomenon referred to by those concepts."[11] The integrated explanation espouses a greater set of relationships between the four perspectives than what has been published in the literature. The key abstract concepts used in this explanation – virtuality, materiality, paradox, management, geography – are such that they can explain coordination in many other concrete GSE events. At the same time, the concepts used have been defined adequately by operations and illustrations in the previous chapters.

ICT connection has been proposed and illustrated (see Chapters 8 and 9) as the idea or explanation for the paradox of GSE coordination. The paradox lies in the theoretical tensions between standardization on the one hand and variation, modularization, and socialization on the other in GSE coordination. The idea of connection exemplifies the introduction of new concepts which Marshall S. Poole and Andrew Van de Ven propose as a strategy for taking advantage of theoretical tensions.[12] The novelty of ICT connection in the theory of GSE coordination lies in how it serves as the explanation for the practical existence of paradox and its theoretical inclusion in coordination theory development. Previous conceptualizations of GSE coordination have maintained the paradox, but researchers have not taken advantage of them to develop more encompassing theories. This is not surprising as we find that previous theories of coordination by plans (standardization, architectures, and structures) have been developed with rationality logic which is very mechanical and linear, aiming for parsimony, precision, and internal consistency. In previous theories of coordination by mutual adjustments, we find that modularization, behavioral discretion, autonomous action, agility, and knowing-in-action reflect indeterminacy logic. We also find that in theories developed with indeterminacy logic that they do not reflect mechanization or standardization. They reflect multidirectional analyses, but they still aim for linearity and consistency which are limiting operative conditions.

The idea of ICT connection which takes advantage of the paradox by encompassing the opposing concepts provides a better field of knowledge because it is wider and more comprehensive than the linear fields in previous GSE theories. ICT connection may be linear in terms of data transmission, but in terms of sensual effect, it provides software engineers with total awareness. Having been identified through a dialectical method, the concept of ICT connection is a knowledge contribution that enables us to analytically separate the two previous sets of concepts. It is also a development based on dynamic, circular, and reciprocal relationships between ICTs, software engineering, geography, and coordination. For this reason, it is not just a nominal concept but an explanatory one. It is meant to shift our attention from its mundane and commonplace character in GSE settings toward its epistemological character. ICT connection, explained as a causal mechanism of a total field of knowledge held by global software engineers for coordinating their work, is an explicit articulation of its causality. In previous research, connectivity as a causal mechanism of coordination has been implied rather than explicitly articulated (see Chapter 8). But this book articulates ICT connection as a phenomenon-specific causal mechanism beyond previous instances where it is implied and almost taken for granted.

Capitalization of technology, information, and human resources is also a concept that enriches our knowledge of previous modes of GSE coordination by mutual adjustments with intelligence and creativity. The notable coordination modes in previous literature are ambidextrous coping,[13] informal communication,[14] knowing-in-practice,[15] and improvisation.[16] These modes are meant to explain how organizations manage task and environmental uncertainties. They enable us to understand how national and socio-cultural diversities among virtual team members are leveraged to address task and environmental uncertainties. But they basically suggest reactive strategies and processes that signify limited conceptualization of GSE coordination. The dialectical interpretation of coordination leads us to transcend adaptation and coping tactics and look at capitalization of technology, information, and human resources. Thus, coordination through capitalization provides us with fresh knowledge of how organizational intelligence combines with ICTs to achieve coordination.

The previous paragraphs challenge us to rethink GSE coordination, which should lead us to discuss how we can challenge ourselves to research and redo GSE coordination henceforth. Our fresh knowledge of virtuality, technology coordination, dataspaces, connection, capitalization, village dependencies, and technology-generated management of village dependencies must offer numerous future research directions and practical guidelines.

The paper entitled "Virtual Teams Research" by Lucy L. Gilson and her colleagues[17] presents a number of constructs classified under inputs, mediators, moderators, and outcomes of virtual teams. Table 10.1 shows several constructs which include theirs. The first future research challenge brought forth by this book is hinged on virtuality as a logical framework or perspective for the study of other virtual team constructs such as those listed in Table 10.1. Virtualization of software teams across geographical and time boundaries is trending considerably in current information systems and software engineering research. Given the independence of digital logic from physical infrastructure, organizations can combine and recombine digital logic into diverse material and virtual forms, functions, and organizational effects. According to Paul M. Leonardi and his colleagues "new technologies bring changes to the way people communicate, act, and organize their social relations."[18] Wherefore, the diverse forms and functions hold potential to generate different technology coordination dynamics in virtual teams. Relating each of these diverse forms and functions to the constructs in Table 10.1 is appropriate because it can lead to a production of new explanations of how virtuality constitutes an insightful logical framework or high-level virtual teamwork assumptions in future research.

Beside virtuality as a proposed logical research framework, the technology explanation also constitutes a proposed analytical framework for future GSE and virtual teams research. A logical framework is a high-level system for understanding the resources, principles, activities, and teleology of some virtual teamwork.

Table 10.1 Constructs for Future Virtual Teams Research

Inputs	Mediators and Moderators		Outputs
• Task	• Communication	• Virtuality	• Performance
• Knowledge	• Collaboration	• Mobility	• Efficiency
• Technology	• Conflict	• Globalization	• Effectiveness
• Leadership	• Trust	• Dependency	• Position
• Composition	• Understanding	• Connectivity	• Innovation
• Culture	• Variation	• Intelligence	• Reputation
• Experience	• Agility	• Absorption	• Resilience
• etc.	• etc.		• etc.

An analytical framework is an explanatory and/or predictive instrument for breaking, sorting, classifying, and interrelating data. Technology coordination as a new theory proposed in this book is an explanatory framework that should be used to study the relationship between virtual teamwork coordination and other constructs. The suggestion here is that researchers, when studying virtual teamwork coordination as the independent or criterion variable, should not fall back on traditional coordination theories. This proposed theory of technology coordination has a distinctive assumption (technology is material), distinctive constructs (connection and capitalization of ICTs, exploitation of geography, paradox of organization, and management of information), and distinctive construct interrelations (connection and capitalization of ICTs, by virtue of technology materiality, leads to organizational paradox, information management, and geography exploitation) (Figure 8.3). The benefits of adopting this analytical framework are integrated analysis, technological interpretation, and potentially new technological explanations of virtual teamwork phenomena.

Extending the frontiers of GSE knowledge also depends more on researching technology coordination as an independent variable than as a dependent variable. When technology coordination is approached from an independent angle, it also becomes a lens for the study of these domains leading to the production of new perspectives on these GSE domains. In particular, if connection and capitalization of ICTs are the primary mechanisms that enable modularization, standardization, and variation to co-exist and function, then further research on how different connections and capitalizations shape different domains is in order.

Based on the proposal of connection and capitalization of ICTs as well as dataspaces, this book also calls for more research on resource and location analytics which are major themes in business intelligence scholarship. Business intelligence is now a burgeoning area in information systems research because of how it facilitates organizational creativity and competitiveness. Given that simulations are at the heart of virtualization, there is already fertile grounds laid in GSE for connection and capitalization of ICTs as well as dataspaces to be used as approaches to the study of other constructs listed in Table 10.1. James G. March[19] calls this exploration of possible new directions for virtual team coordination as opposed to exploitation which refers to its refinement. The concept of dataspace is especially worth exploring in relation to constructs in resource and locational analytics.

The wealth of ideas generated from this exploration in turn present further opportunities for exploitation of dataspaces to develop its different and subsidiary strands in different GSE configurations in order to deepen our understanding of coordination. An important question, for instance, is how are different dataspaces in GSE configurations implicated in coordination? Further research also needs be undertaken to explore the question, how are different strands of GSE coordination achieved under different ICT connections and capitalizations?

Alongside the pursuit to address these questions, the concept of village dependencies (derived from cosmic consciousness and total field awareness that are media ecological concepts) also needs to be exploited in future research. Different GSE configurations and different electrical media deployed therein will generate different village dependencies. As a corollary, there will be different combinations of technology and human resources leveraged to manage the different village dependencies. Therefore, there are fertile grounds for further research to produce a taxonomy or refinement of village dependencies based on configuration and media (type I theory in Shirley Gregor's ontology[20]). As taxonomies lack explanations and predictions, the strands of village dependencies in the taxonomy must be subjects of further exploration to explain and/or predict how and why they generate different degrees of GSE coordination (Gregor's types II to IV theories). This book has not been able to achieve these explorations and exploitations because it has largely focused on how these concepts, in their unrefined states, shape coordination through an idiographic study. The concepts need refinement and the refinements should be explored in terms of how and why they are implicated in the listed constructs in Table 10.1.

Our next discussions are about how we should redo GSE coordination in light of the contributions made by this book.

Technology coordination of GSE as a new epistemology of coordination (arising from this book's technological explanation) emphasizes the essential role of ICT ahead of its instrumental role. This implies that ICT organizations that establish and use GSE teams should strategize rather than just commoditize ICT resource deployment and management to achieve coordination. ICT resource strategy enables optimal use of current resources and intelligent planning of future resources through recognition of trends and patterns. It also enables the tactical implementation of ICT resource strategic plans by watching for contingencies in ICT use and coordination. Organizations should also set up systems that can gather information on the relationship between software resources and contingencies that cause the exploitation of the resources. Organizations can then use ICT resources to process the information through resource and location analytics in order to increase the organization's intelligence capacity for capitalization.

The theory of coordination by ICTs, as the absolute idea of managing dependencies, uncertainties, and conflicts in GSE, means that project managers should prioritize ICT connection and capitalization modes of coordination. This translates into investments in rich digital communication media that enhance co-presence of spatially distributed software developers. It also suggests that criteria for hiring software developers should not be just expertise but also dexterity (dynamic creative capabilities). For example, there is wisdom in investments in communication media with high-definition, widescreen, and interactive monitors that can display real-time simulations of future scenarios and representation of remote situations for coordination purposes. This investment underscores the positioning of these technologies of virtualization as generative mechanisms of GSE coordination. There is also wisdom in investing in developers who have dynamic creative capabilities to exploit technology, information, and human resources ahead of emergent uncertainties.

The proposition of virtuality as the primary logical framework for GSE coordination in this book (ahead of rationality and indeterminacy) suggests to organizations the need to also prioritize software engineers' creativity ahead of their efficiency and effectiveness. Prioritization of creativity is expressed practically through deployment of diverse technological and communication media that enables diverse interaction modes across sites. Software engineers use these media and their enablers to capitalize on ICT in order to practice emergent coordination. We have learned that GSE tasks are discontinuous and result in the emergence of unexpected task and environmental challenges that do not have a match in the existing variation profile of a GSE team (Chapter 9). Therefore, ICT deployment by GSE organizations must have adequate variation that would enable software engineers to capitalize on the ICTs to address emergencies in a creative manner.

The reality of village dependencies generated by electrical ICT media must become a conscious and deliberate design by GSE organizations. This suggestion is sensible in the light of the proposition that village dependencies have technological origins. Moreover, village dependencies are replicates of collocated software engineering dependencies. Their technological origins and virtual logic together have direct implications for design of GSE teams. Recall from Chapter 4 that village dependencies have a virtual logic which aligns with village which is the solution, as well as a rational logic which aligns with global which is the problem. Therefore, deliberate and conscious design of village dependencies provide GSE organizations with control over problem and solution dimensions of coordination. They should use ICTs to make village dependencies both a problem and solution design so that they can manage the dependencies more effectively.

The knowledge of dataspaces (digital information spaces), and their utilization for GSE coordination, suggests that organizations have greater opportunities and scope to manage village dependencies compared with opportunities and scope provided by traditional dependencies. Dataspaces constitute a spatial resource that can be generated with ICTs and manipulated to manage village dependencies. Dataspaces can be isolated from workplaces and then digital representations of software engineers and their work can be manipulated in multiple ways depending on the dynamics of the task and environmental challenges. Thus, the organization does not need to encumber itself with manipulating the physical ICT infrastructure if it wants to manipulate its dataspace to achieve coordination. This is especially recommended in instances where the focus of coordination is at the task rather than the component level. At the task level, designing high cross-site dependencies is critical, and utilization and appropriation of dataspaces are practical steps toward the achievement of this design.

Appendix

Table A.1 Summarized Review of Literature on Coordination in GSE

Problem Domain	References	Unit of Analysis	Key Construct in Focus
Processes	Srikanth & Puranam (2011, 2014)	Knowledge	Dependency
Development speed and communication delays	Herbsleb & Mockus (2003); Herbsleb et al. (2000)	Task	
Evolution	Sabherwal (2003)	Mechanisms	
Distributed knowledge	Ovaska et al. (2003)	Architecture	
Unresolved dependencies	Lindberg et al. (2016)	Implementation and knowledge	
Business process	D'Aubeterre et al. (2008)	Security requirements	
Adaptation of collaboration technology	Thomas & Bostrom (2010)	Team leader strategies	
Task execution	Mani et al. (2014)	Modularization and information sharing	
Knowledge transfer	Kotlarsky et al. (2007)	Knowledge and communication	
Structural impediments	David et al. (2008)	Collaboration	
Development speed and quality	Colazo & Fang (2010)	Temporal dispersion	
Team performance	Espinosa et al. (2007a)	Familiarity, task	

(Continued)

Table A.1 (Continued)   Summarized Review of Literature on Coordination in GSE

Problem Domain	References	Unit of Analysis	Key Construct in Focus
Engineering process	Aversano et al. (2004)	Workflow management technology	
Development management	Paasivaara & Lassenius (2003)	Collaboration practices	
Robust teamwork	Im et al. (2005)	Communication genres	
Knowledge transfer	Koppman & Gupta (2014)	Mutual knowledge	
Time-cost modeling	Espinosa & Carmel (2003, 2004); Espinosa et al. (2012)	Temporal distance	
Knowledge sharing	Espinosa et al. (2007b); Kotlarsky & Oshri (2005); Kotlarsky et al. (2008); Zahedi et al. (2016)	Knowledge	Uncertainty
Knowledge sharing/ transfer	Oshri et al. (2008)	Transactive memory	
Knowledge sharing and performance	Du et al. (2011)	Culture and trust	
Adaptation of collaboration technology	Thomas & Bostrom (2010)	Team leader strategies	
Structural impediments	David et al. (2008)	Collaboration	
Knowledge transfer	Mattarelli & Gupta (2009)	Sub-group dynamics	
Plans and processes	Herbsleb & Grinter (1999); Lee et al. (2013)	Task, communication	
Cultural dynamics	Niederman & Tan (2011)	Culture	Conflict
Knowledge transfer	Zimmermann & Ravishankar (2014)	Social capital	
Robust teamwork	Im et al. (2005)	Communication genres	

Table A.2 Summarized Review of the Literature on Electronic Meetings and on GSE

	Key Parameter	References	Limitations
GSE research on the functionality of teleconferences	Task complexity	Dennis et al. (2001); Zigurs & Buckland (1998)	They do not give specific attention to coordination, much more to the challenges induced by software complexity and global distribution of resources.
	Social action	Ngwenyama (1998)	
	Interdependence construction	Karsten (2003)	
	Strategic management	Tyran et al. (1992)	
	Conflict management	Poole et al. (1991); Sambamurthy & Poole (1992)	
	Team building	Esbensen & Bjørn (2014)	
	Standardization and routine in dependencies	Huang et al. (2003)	
	Agile methods	Bjørn et al. (2019)	
	Knowledge sharing	Zahedi et al. (2016)	
	Group structure	McLeod & Liker (1992)	

(Continued)

Table A.2 (Continued)   Summarized Review of the Literature on Electronic Meetings and on GSE

	Key Parameter	References	Limitations
GSE research on coordination	Obtaining the right balance between formal and informal communications	Grinter et al. (1999)	Do not implicate teleconferences directly
	Knowledge sharing	Ghobadi (2015)	
	Using software architectures, plans and informal ad hoc communications	Herbsleb & Grinter (1999)	
	Mechanisms and processes that can address distance-related delays in communications	Herbsleb & Mockus (2003)	Do not particularly address the potential role of teleconferences as coordination processes
	Knowledge-based perspective: facilitating knowledge flows, making knowledge explicit, amplifying knowledge, and building social capital	Kotlarsky et al. (2008)	
	Global dispersion and task complexity	Lee et al. (2013)	
	Global dispersion	Nguyen-Duc et al. (2015)	
	Time/cost modeling of the effects of distance	Espinosa & Carmel (2003; 2007b)	

Notes

Chapter 1

1. Overby, E. (2008). Process Virtualization Theory and the Impact of Information Technology. *Organization Science, 19*(2), 277–291.
2. Kraut, R. E., & Streeter, L. A. (1995). Coordination in Software Development. *Communications of the ACM, 38*(3), 69–81.

Chapter 2

1. Kraut, R. E., & Streeter, L. A. (1995). Coordination in Software Development. *Communications of the ACM, 38*(3), 69–81.
   Crowston, K. (1997). A Coordination Theory Approach to Organizational Process Design. *Organization Science, 8*(2), 157–175.
2. Thompson, J. D. (2003/1967). *Organizations in Action* (2nd ed.). New Brunswick, NJ: Transactions Publishers.
   March, J. G., & Simon, H. A. (1993/1958). *Organizations* (2nd ed.). Cambridge, MA: Blackwell.
   Van de Ven, A. H., Delbecq, A. L., & Koenig, R. (1976). Determinants of Coordination Modes within Organizations. *American Sociological Review, 41*(2), 322–328.
   Mintzberg, H. (1983). *Structure in Fives: Designing Effective Organizations*. Englewood Cliffs, NJ: Prentice-Hall.
   Quinn, R. W., & Dutton, J. E. (2005). Coordination as Energy-In-Conversation. *Academy of Management Review, 30*(1), 36–57.
3. Daft, R. L., & Lengel, R. H. (1986). Organizational Information Requirements, Media Richness and Structural Design. *Management Science, 32*(5), 554–571.
   Milliken, F. J. (1987). Three Types of Perceived Uncertainty about the Environment: State, Effect, and Response Uncertainty. *Academy of Management Review, 12*, 133–143.
4. Schmidt, S. M., & Kochan, T. A. (1972). Conflict: Toward Conceptual Clarity. *Administrative Science Quarterly, 17*(3), 359–370.

Victor, B., & Blackburn, R. S. (1987). Interdependence: An Alternative Conceptualisation. *Academy of Management Review, 12*(3), 486–498.

Camerer, C., & Knez, M. (1996). Coordination, Organizational Boundaries and Fads in Business Practices. *Industrial and Corporate Change, 5*(1), 89–112.

Montoya-Weiss, M. M., Massey, A. P., & Song, M. (2001). Getting it Together: Temporal Coordination and Conflict Management in Global Virtual Teams. *Academy of Management Journal, 44*(6), 1251–1262.

5. Lawrence, P. R., & Lorsch, J. W. (1967a). Differentiation and Integration in Complex Organizations. *Administrative Science Quarterly, 12*(1), 1–47.

   Lawrence, P. R., & Lorsch, J. W. (1967b). *Organization and Environment.* Homewood, IL: Richard D. Irwin.

6. Van de Ven, A. H., & Delbecq, A. L. (1974). A Task Contingent Model of Work-Unit Structure. *Administrative Science Quarterly, 19*(2), 183–197.

   Van de Ven et al., *ibid.*

7. Weick, K. E. (1979). *The Social Psychology of Organizing.* New York: McGraw Hill.

   Belanger, F., & Collins, R. W. (1998). Distributed Work Arrangements: A Research Framework. *The Information Society, 14*(2), 137–152.

8. Weick, *ibid.*

9. Cook, S. D. N., & Brown, J. S. (1999). Bridging Epistemologies: The Generative Dance between Organizational Knowledge and Organizational Knowing. *Organization Science, 10*(4), 381–400.

10. Bateson, G. (1979). *Mind and Nature: A Necessary Unity.* New York: E. P. Dutton.

11. Daft & Lengel, *ibid.*, p. 565.

12. Victor & Blackburn, *ibid.*

    Milliken, *ibid.*

    McCann, J. E., & Ferry, D. L. (1979). An approach for Assessing and Managing Inter-unit Interdependence. *Academy of Management Review, 4*(1), 113–119.

    Duncan, R. B. (1972). Characteristics of Organizational Environments and Perceived Environmental Uncertainty. *Administrative Science Quarterly, 17*(3), 313–327.

    Leifer, R., & Huber, G. P. (1977). Relations among Perceived Environmental Uncertainty, Organizational Structure, and Boundary-Spanning Behavior. *Administrative Science Quarterly, 22*(2), 235–247.

    Argote, L. (1982). Input Uncertainty and Organizational Coordination in Hospital Emergency Units. *Administrative Science Quarterly, 27*(3), 420–434.

13. Milliken, *ibid.*, p. 136.

14. Van de Ven et al., *ibid.*

15. Gioia, D. A. (1986). Symbols, Scripts, and Sensemaking: Creating Meaning in the Organizational Experience. In H. P. Sims & D. A. Gioia (Eds.), *The Thinking Organization.* San Francisco, CA: Jossey-Bass, pp. 49–74.

16. Daft & Lengel, *ibid.*

17. Jehn, K. A. (1997). A Qualitative Analysis of Conflict Types and Dimensions in Organizational Groups. *Administrative Science Quarterly, 42*, 530–557.

18. Thompson, J. D. (1960). Organizational Management of Conflict. *Administrative Science Quarterly, 4*, 389–409.

19. Camerer & Knez, *ibid.*

20. Kretschmer, T., & Puranam, P. (2008). Integration through Incentives in Differentiated Organizations. *Organization Science, 19*, 860–875.

21. Nidumolu, S. R. (2001). The Effect of Coordination and Uncertainty on Software Project Performance: Residual Performance Risk as an Intervening Variable. *Information Systems Research, 6*(3), 191–219.
    Kraut & Streeter, *ibid.*
22. Lee, G., Espinosa, J. A., & DeLone, W. H. (2013). Task Environment Complexity, Global Team Dispersion, Process Capabilities, and Coordination in Software Development. *IEEE Transactions on Software Engineering, 39*(12), 1753–1769.
    Srikanth, K., & Puranam, P. (2014). The Firm as a Coordination System: Evidence from Software Services Offshoring. *Organization Science, 25*(4), 1253–1271.
    Niederman, F., & Tan, F. B. (2011). Managing Global IT Teams: Considering Cultural Dynamics. *Communications of the ACM, 54*(4), 24–27.
    Cataldo, M., & Herbsleb, J. D. (2013). Coordination Breakdowns and Their Impact on Development Productivity and Software Failures. *IEEE Transactions of Software Engineering, 30*(3), 343–360.
23. Conway, M. E. (1968). How do Committees Invent? *Datamation, 14*(5), 28–31.
24. Brooks, F. P. (1987). No Silver Bullet: Essence and Accidents of Software Engineering. *IEEE Computer, 20*(4), 10–19.
25. Herbsleb, J. D., Mockus, A., Finholt, T. A., & Grinter, R. E. (2000). *Distance, Dependencies, and Delay in Global Collaboration*. Paper presented at the Conference on Computer Supported Cooperative Work, Philadelphia, PA.
    Sahay, S., Nicholson, B., & Krishna, S. (2003). *Global IT Outsourcing: Software Development Across Borders*. Cambridge: Cambridge University Press.
26. Brooks, *ibid.*
27. Espinosa, J. A., Slaughter, S. A., Kraut, R. E., & Herbsleb, J. D. (2007). Team Knowledge and Coordination in Geographically Distributed Software Development. *Journal of Management Information Systems, 24*(1), 135–169.
    Kotlarsky, J., & Oshri, I. (2005). Social Ties, Knowledge Sharing and Successful Collaboration in Globally Distributed System Development Projects. *European Journal of Information Systems, 14*, 37–48.
    Srikanth, K., & Puranam, P. (2011). Integrating Distributed Work: Comparing Task Design, Communication, and Tacit Coordination Mechanisms. *Strategic Management Journal, 32*(8), 849–875.
    Srikanth, K., & Puranam, P. (2014). The Firm as a Coordination System: Evidence from Software Services Offshoring. *Organization Science, 25*(4), 1253–1271.
28. Niederman & Tan, *ibid.*
29. Herbsleb, J. D., & Grinter, R. E. (1999). Architectures, Coordination, and Distance: Conway's Law and Beyond. *IEEE Software, 16*(5), 63–70.
    Herbsleb, J. D., & Mockus, A. (2003). An Emprical Study of Speed and Communication in Globally Distributed Software Development. *IEEE Transactions on Software Engineering, 29*(6), 481–494.
    Herbsleb, J. D., Mockus, A., Finholt, T. A., & Grinter, R. E. (2000). *Distance, Dependencies, and Delay in Global Collaboration*. Paper presented at the Conference on Computer Supported Cooperative Work, Philadelphia, PA.
    Lee et al., *ibid.*
30. Espinosa, J. A., & Carmel, E. (2003). The Impact of Time Separation on Coordination in Global Software Teams: A Conceptual Foundation. *Software Process Improvement and Practice, 8*, 249–266.

Espinosa, J. A., & Carmel, E. (2004). *The Effect of Time Separation on Coordination Costs in Global Software Teams: A Dyad Model.* . Paper presented at the 37th Hawaiian International Conference on System Sciences, Big Island, HI.

Espinosa, J. A., Cummings, J. N., & Pickering, C. (2012). Time Separation, Coordination, and Performance in Technical Teams. *IEEE Transactions on Engineering Management, 59*(1), 91–103.

31. Ovaska, P. A., Rossi, M., & Marttiin, P. (2003). Architecture as a Coordination Tool in Multi-site Software Development. *Software Process Improvement and Practice, 8*, 233–247.

32. Yates, J., & Orlikowski, W. J. (1992). Genres of Organizational Communication: A Structurational Approach to Studying Communication and Media. *Academy of Management Review, 17*(2), 299–326.

33. Galbraith, J. R. (1974). Organization Design: An Information Processing View. *Interfaces, 4*(3), 28–36.

34. Tushman, M. L., & Nadler, D. A. (1978). Information Processing as an Integrating Concept in Organizational Design. *Academy of Management Review, 3*(3), 613–624.

35. Sabherwal, R. (2003). The Evolution of Coordination in Outsourced Software Development Projects: A Comparison of Client and Vendor Perspectives. *Information and Organization, 13*, 153–202.

36. Herbsleb & Grinter, *ibid*.

Herbsleb & Mockus, *ibid*.

Herbsleb, J. D., Mockus, A., Finholt, T., & Grinter, R. E. (2001). *An Empirical Study of Global Software Development: Distance and Speed.* Paper presented at the 23rd International Conference on Software Engineering, Ontario, Canada.

Herbsleb, J. D., Mockus, A., Finholt, T. A., & Grinter, R. E. (2000). *Distance, Dependencies, and Delay in Global Collaboration.* Paper presented at the Conference on Computer Supported Cooperative Work, Philadelphia, PA.

37. Grinter, R. E., Herbsleb, J. D., & Perry, D. E. (1999). *The Geography of Coordination: Dealing with Distance in R&D Work.* Paper presented at the GROUP'99, Phoenix, AZ.

38. Huang, H., & Trauth, E. M. (2008). *Cultural Influences on Temporal Separation and Coordination in Globally Distributed Software Development.* Paper presented at the 29th International Conference on Information Systems, Paris, France.

39. Kotlarsky, J., Fenema, P. C. v., & Willcocks, L. P. (2008). Developing a Knowledge-Based Perspective on Coordination: The Case of Global Software Projects. *Information and Management, 45*, 96–108.

40. Carmel, E., & Abbott, P. (2007). Why 'Nearshore' Means That Distance Matters. *Communications of the ACM, 50*(10), 40–46.

Carmel, E., & Argarwal, R. (2001). Tactical Approaches for Alleviating Distance in Global Software Development. *IEEE Software, 18*(2), 22–29.

41. D'Mello, M., & Sahay, S. (2007). "I Am Kind of a Nomad Where I Have to Go Places and Places" ... Understanding Mobility, Place and Identity in Global Software Work from India. *Information and Organization, 17*(3), 162–192.

42. Dourish, P. (2006). *Re-Space-ing Place: "Place" and "Space" Ten Years On.* Paper presented at the Computer Supported Cooperative Work (CSCW'06) Conference, Alberta, Canada.

Schultze, U., & Boland, R. J. (2000). Place, Space and Knowledge Work: A Study of Outsourced Computer System Administrators. *Accounting, Management and Information Technology, 10*, 187–219.

43. Agnew, J. (2011). Space and Place. In J. Agnew & D. N. Livingstone (Eds.), *The SAGE Handbook of Geographical Knowledge*. London: Sage, pp. 316–330.
44. Agerfalk, P., Fitzgerald, B., Holmström, H., Lings, B., Lundell, B., & O Conchuir, E. (2005). *A Framework for Considering Opportunities and Threats in Distributed Software Development*. Paper presented at the International Workshop on Distributed Software Development (DiSD'05), Paris, France.
45. Ramasubbu, N., Cataldo, M., Balan, R. K., & Herbsleb, J. D. (2011). *Configuring Global Software Teams: A Multi-Company Analysis of Project Productivity, Quality, and Profits*. Paper presented at the International Conference on Software Development, Honolulu, HI.
46. Maruping, L. M., Zhang, X., & Venkatesh, V. (2009). Role of Collective Ownership and Coding Standards in Coordinating Expertise in Software Project Teams. *European Journal of Information Systems, 18*(4), 355–371.
47. Ngwenyama, O. (2009). Virtual Team Collaboration: Building Shared Meaning, Resolving Breakdowns and Creating Translucence. *Information Systems Journal, 19*(3), 227–253.
48. Espinosa, J. A., & Carmel, E. (2003). The Impact of Time Separation on Coordination in Global Software Teams: A Conceptual Foundation. *Software Process Improvement and Practice, 8*, 249–266.
  Hinds, P. J., & Bailey, D. E. (2003). Out of Sight, Out of Sync: Understanding Conflict in Distributed Teams. *Organization Science, 14*(6), 615–632.
  Maznevski, M. L., & Choduba, K. M. (2000). Bridging Space over Time: Global Virtual Team Dynamics and Effectiveness. *Organization Science, 11*(5), 473–492.
  Montoya-Weiss et al., *ibid*.
49. Faraj, S., & Sproull, L. (2000). Coordinating Expertise in Software Development Teams. *Management Science, 46*(12), 1554–1568.
50. Zolin, R., Hinds, P. J., Fruchter, R., & Levitt, R. E. (2004). Interpersonal Trust in Cross-Functional Geographically Distributed Work. A Longitudinal Study. *Information and Organization, 14*, 1–26.
51. Ovaska et al., *ibid*.
52. Ribes, D., Jackson, S., Geiger, S., Burtond, M., & Finholt, T. (2013). Artifacts that organize: Delegation in the distributed organization. *Information and Organization, 23*(1), 1–14.
53. Sabherwal, *ibid*.
54. Vlaar, P. W. L., Fenema, P. C. V., & Tiwari, V. (2008). Cocreating Understanding and Value in Distributed Work: How Members of Onsite and Offshore Vendor Teams Give, Make, Demand, and Break Sense. *MIS Quarterly, 32*(2), 227–255.
55. Smith, W. K., & Lewis, M. W. (2011). Toward A Theory of Paradox: A Dynamic Equilibrium Model of Organizing. *Academy of Management Review, 36*(2), 381–403.
56. Scott, W. R. (2003). *Organizations: Rational, Natural and Open Systems* (5th ed.). Upper Saddle River, NJ: Prentice Hall.
57. Malhotra, A., & Majchrzak, A. (2012). How Virtual Teams Use Their Virtual Workspace to Coordinate Knowledge. *ACM Transactions on Management Information Systems, 3*(1), 1–14.
58. Leonardi, P. M., Nardi, B. M., & Kallinikos, J. (Eds.). (2012). *Materiality and Organizing: Social Interaction in a Technological World*. Oxford: Oxford University Press.
59. Mingers, J., Mutch, A., & Willcocks, L. P. (2013). Critical Realism in Information Systems Research. *MIS Quarterly, 37*(3), 795–802.

60. Dennis, A. R., Fuller, R. M., & Valacich, J. S. (2008). Media, Tasks, and Communication Processes: A Theory of Media Synchronicity. *MIS Quarterly, 32*(3), 575–600.
61. Galbraith, J. R. (1974). Organization Design: An Information Processing View. *Interfaces, 4*(3), 28–36.
    Galbraith, J. R. (1977). *Organization Design*. Reading, MA: Addison-Wesley.
62. Tushman & Nadler, *ibid.*
63. Daft & Lengel, *ibid.*
64. Daft, R. L., & Weick, K. E. (1984). Toward a Model of Organizations as Interpretation Systems. *Academy of Management Review, 9*(2), 284–295.
65. Daft & Weick, *ibid.*
66. Borgmann, A. (1984). *Technology and the Character of Contemporary Life*. London: University of Chicago Press.
67. Audretsh, D. B. (1998). Agglomeration and the Location of Innovative Activity. *Oxford Review of Economic Policy, 14*(2), 18–29.
68. Cairncross, F. (1997). *The Death of Distance: How the Communications Revolution will Change Our Lives*. Boston, MA: Harvard Business School Press.
69. Feldman, M. P. (2000). Location and Innovation: The New Economic Geography of Innovation, Spillovers and Agglomeration. In G. L. Clark, M. P. Feldman, & M. S. Gertler (Eds.), *The Oxford Handbook of Economic Geography*. Oxford: Oxford University Press, pp. 373–395.
70. Hetherington, K. (1998). In Place of Geometry: The Materiality of Place. *The Sociological Review, 45*(1), 183–199.
71. Glassman, R. B. (1973). Persistence and Loose Coupling in Living Systems. *Behavioral Science, 18*(2), 83–98.
    Orton, J. D., & Weick, K. E. (1990). Loosely Coupled Systems: A Reconceptualization. *Academy of Management Review, 15*(2), 203–223.
72. Hegel, G. W. F. (1977/1807). *The Phenomenology of Spirit*. Oxford: Oxford University Press.

Chapter 3

1. Overby, E. (2008). Process Virtualization Theory and the Impact of Information Technology. *Organization Science, 19*(2), 277–291.
2. D'Mello, M., & Eriksen, T. H. (2010). Software, Sports Day and Sheera: Culture and Identity Processes within a Global Software Organization in India. *Information and Organization, 20*(2), 81–110.
   Herbsleb, J. D., & Grinter, R. E. (1999). Architectures, Coordination, and Distance: Conway's Law and Beyond. *IEEE Software, 16*(5), 63–70.
   Kotlarsky, J., Fenema, P. C. V., & Willcocks, L. P. (2008). Developing a Knowledge-Based Perspective on Coordination: The Case of Global Software Projects. *Information and Management, 45*, 96–108.
3. D'Mello & Eriksen, *ibid.*
   D'Mello, M., & Sahay, S. (2007). "I Am Kind of a Nomad Where I Have to Go Places and Places" ... Understanding Mobility, Place and Identity in Global Software Work from India. *Information and Organization, 17*(3), 162–192.

Huang, W., & Trauth, E. M. (2008). *Cultural Influences on Temporal Separation and Coordination in Globally Distributed Software Development.* Paper presented at the International Conference on Information Systems, Paris, France.

4. Kallinikos, J. (2012). Form, Function, and Matter: Crossing the Border of Materiality. In P. M. Leonardi, B. M. Nardi, & J. Kallinikos (Eds.), *Materiality and Organizing: Social Interaction in a Technological World.* Oxford: Oxford University Press, pp. 67–87.

5. Kallinikos, J. (2006). Information out of Information: On the Self-Referential Dynamics of Information Growth. *Information Technology and People, 19*(1), 98–115.

6. Tilson, D., Lyytinen, K., & Sørensen, C. (2010). Digital Infrastructures: The Missing IS Research Agenda. *Information Systems Research, 21*(4), 748–759.

7. Bailey, D. E., Leonardi, P. M., & Barley, S. R. (2012). The Lure of the Virtual. *Organization Science, 23*(5), 1485–1504.

8. Scott, W. R. (2003). *Organizations: Rational, Natural and Open Systems* (5th ed.). Upper Saddle River, NJ: Prentice Hall, p. 33.

9. Barnard, C. I. (1938). *The Functions of the Executive.* Cambridge, MA: Harvard University Press.
   Mayo, E. (1933/2003). *The Human Problems of Industrial Civilization.* New York: Routledge.
   Roethlisberger, F. J., & Dickson, W. J. (1939). *Management and the Worker.* Cambridge, MA: Harvard University Press.

10. Stinchcombe, A. L. (1990). *Information and Organizations.* Berkeley, CA: University of California Press.

11. von Bertalanffy, L. (1969). *General System Theory: Foundations, Development, Applications.* New York: George Braziller.

12. Schmidt, K., & Simone, C. (1996). Coordination Mechanisms: Towards a Conceptual Foundation of Computer Supported Cooperative Work Systems Design. *Computer Supported Cooperative Work: The Journal of Collaborative Computing, 5*, 155–200.

13. Van de Ven, A. H., Delbecq, A. L., & Koenig, R. (1976). Determinants of Coordination Modes within Organizations. *American Sociological Review, 41*(2), 322–328.

14. March, J. G., & Simon, H. A. (1993/1958). *Organizations* (2nd ed.). Cambridge, MA: Blackwell.

15. Thompson, J. D. (2003/1967). *Organizations in Action* (2nd ed.). New Brunswick, NJ: Transactions Publishers (First published in 1967).

16. Malone, T. W., & Crowston, K. (1990). *What is Coordination Theory and How Can It Help Design Cooperative Work Systems?* Paper presented at the 3rd Conference on Computer-Supported Cooperative Work, New York.
   Malone, T. W., & Crowston, K. (1994). The Interdisciplinary Study of Coordination. *ACM Computing Surveys, 26*(1), 87–119.

17. Quinn, R. W., & Dutton, J. E. (2005). Coordination as Energy-In-Conversation. *Academy of Management Review, 30*(1), 36–57.

18. Kogut, B., & Zander, U. (1996). What Firms Do? Coordination, Learning and Identity. *Organization Science, 7*(5), 502–518.

19. Gittell, J. H. (2011). New Directions for Relational Coordination Theory. In K. S. Cameron & G. Spreitzer (Eds.), *Oxford Handbook of Positive Organizational Scholarship.* Oxford: Oxford University Press, pp. 74–94.

20. Weick, K. E. (2009). *Making Sense of the Organization: The Impermanent Organization: Volume 2.* Chichester: John Wiley & Sons.

21. Faraj, S., & Sproull, L. (2000). Coordinating Expertise in Software Development Teams. *Management Science, 46*(12), 1554–1568.
22. Bechky, B. A. (2006). Gaffers, Gofers and Grips: Role-based Coordination in Temporary Organizations. *Organization Science, 17*(1), 3–21.
23. Valentine, M. A., & Edmondson, A. C. (2014). Team Scaffolds: How Mesolevel Structures Enable Role-Based Coordination in Temporary Groups. *Organization Science, 26*(2), 405–422.
24. Heckscher, C., & Adler, P. S. (2006). *The Firm As A Collaborative Community: Reconstructing Trust in the Knowledge Economy.* Oxford: Oxford University Press.
25. Daft, R. L., & Lengel, R. H. (1986). Organizational Information Requirements, Media Richness and Structural Design. *Management Science, 32*(5), 554–571.
26. Perrow, C. (1967). A Framework for the Comparative Analysis of Organizations. *American Sociological Review, 32*(2), 194–208.
27. Campbell, D. J. (1988). Task Complexity: A Review and Analysis. *Academy of Management Review, 13*(1), 40–52.
28. Van de Ven et al., *ibid.*
29. Mathiassen, L., & Stage, J. (1992). The Principle of Limited Reduction in Software Design. *Information Technology & People, 6*(2–3), 171–185.
30. Campbell, *ibid.*
31. Thompson, *ibid.*
32. Watson-Manheim, M. B., Chudoba, K. M., & Crowston, K. (2002). Discontinuities and Continuities: A New Way to Understand Distributed Work. *Information Technology and People, 15*(3), 191–209.
33. Lee, G., Delone, W., & Espinosa, J. A. (2006). Ambidexterous Coping Strategies in Globally Distributed Software Development Projects. *Communications of the ACM, 49*(10), 35–40.
34. McLuhan, M. (1964/2001). *Understanding Media: The Extensions of Man.* London: Routledge.
35. McLuhan, *ibid.,* p. 64.
36. Kallinikos, J. (2006). Information out of Information: On the Self-Referential Dynamics of Information Growth. *Information Technology and People, 19*(1), 98–115.
37. McLuhan, *ibid.,* p. 385.
38. McLuhan, *ibid.*
39. Stratte, L. (2008). Studying Media as Media. *Media Tropes, 1,* 127–142.
40. Culkin, J. (1967). Each Culture Develops Its Own Sense Ratio to Meet the Demands of its Environment. In G. Stearn (Ed.), *McLuhan: Hot and Cool.* New York: New American Library, pp. 49–57.
41. Leonardi, P. M., & Barley, S. R. (2008). Materiality and Change: Challenges to Building Better Theory about Technology and Organizing. *Information and Organization, 18,* 159–176.
42. Bailey et al., *ibid.*
43. Borgmann, A. (1999). *Holding on to Reality: The Nature of Information at the Turn of the Century.* Chicago, IL: University of Chicago Press.
44. Bailey et al., *ibid.*
45. Bailey et al., *ibid.*
46. Kraut, R. E., Steinfield, C., Chan, A. P., Butler, B., & Hoag, A. (1999). Coordination and Virtualization: The Role of Electronic Networks and Personal Relationships. *Organization Science, 10*(6), 722–740.

Chapter 4

1. Short, J., Williams, E., & Christie, B. (1976). *The Social Psychology of Telecommunications*. New York: John Wiley.
2. Dashiell, J. F. (1935). Experimental Studies of the Influence of Social Situations on the Behavior of Individual Human Adults. In C. Murchison (Ed.), *Handbook of Social Psychology*. Worcester, MA: Clark University Press, pp. 1097–1158.
3. Tushman, M. L. (1979). Work Characteristics and Subunit Communication Structure: A Contingency Analysis. *Administrative Science Quarterly, 24*(1), 82–98.
4. Perrow, C. (1967). A Framework for the Comparative Analysis of Organizations. *American Sociological Review, 32*(2), 194–208. The work categories are non-routine, engineering, craft, and routine, defined by task analyzability and number of exceptions. Non-routine tasks have a high number of exceptions and are unanalyzable, and routine has opposite characteristics. Engineering tasks are analyzable but have many exceptions. Craft tasks are unanalyzable but have few exceptions.
5. Daft, R. L., & Lengel, R. H. (1984). Information Richness: A New Approach to Manager Information Processing and Organizational Design. In B. Straw & L. L. Cummings (Eds.), *Research in Organizational Behavior*. Greenwich, CT: JAI Press, pp. 191–233.
   Daft, R. L., & Lengel, R. H. (1986). Organizational Information Requirements, Media Richness and Structural Design. *Management Science, 32*(5), 554–571.
6. Overby, E. (2008). Process Virtualization Theory and the Impact of Information Technology. *Organization Science, 19*(2), 277–291.
7. Winograd, T. (1987/1988). A Language/Action Perspective on the Design of Cooperative Work. *Human-Computer Interaction, 3*, 3–30.
8. Searle, J. R. (1969). *Speech Acts: An Essay in the Philosophy of Language*. New York: Cambridge University Press.
9. Dennis, A. R., Fuller, R. M., & Valacich, J. S. (2008). Media, Tasks, and Communication Processes: A Theory of Media Synchronicity. *MIS Quarterly, 32*(3), 575–600.
10. Dennis et al., *ibid.,* p. 575.
11. Dennis, A. R., Aronson, J. E., Heninger, W. E., & Walker, E. D. (1999). Structuring Time and Task in Electronic Brainstorming. *MIS Quarterly, 23*(1), 95–108.
12. Dennis, A. R., & Valacich, J. S. (1993). Computer Brainstorms: More Heads Are Better than One. *Journal of Applied Psychology, 78*, 531–537.
13. Cramton, C. D. (2001). The Mutual Knowledge Problem and Its Consequences for Dispersed Collaboration. *Organization Science, 12*(3), 346–371.
    Grosse, C. U. (2002). Managing Communication within Virtual Intercultural Teams. *Business Communication Quarterly, 65*(4), 22–38.
    Kimble, C. (2011). Building Effective Virtual Teams: How to Overcome the Problems of Trust and Identity in Virtual Teams. *Global Business and Organizational Excellence, 30*(2), 6–15.
14. McLuhan, M. (1964). *Understanding Media: The Extensions of Man*. London: Routledge.
15. Strate, L. (2017). *Media Ecology: An Approach to Understanding the Human Condition*. New York: Peter Lang.
16. Leonardi, P. M., & Barley, S. R. (2008). Materiality and Change: Challenges to Building Better Theory about Technology and Organizing. *Information and Organization, 18*, 159–176.

17. Giddens, A. (1984). *The Constitution of Society: Outline of the Theory of Structure.* Berkeley, CA: University of California Press.
18. Leonardi, P. M. (2012). Materiality, Sociomateriality, and Socio-Technical Systems: What Do These Terms Mean? How are They Different? Do We Need Them?. In P. M. Leonardi, B. M. Nardi, & J. Kallinikos (Eds.), *Materiality and Organizing: Social Interaction in a Technological World.* Oxford: Oxford University Press, pp. 25–48.
19. Orlikowski, W. J. (2007). Sociomaterial Practices: Exploring Technology at Work. *Organization Studies, 28*(9), 1435–1448.
    Orlikowski, W. J., & Scott, S. V. (2008). Sociomateriality: Challenging the Separation of Technology, Work and Organization. *Academy of Management Annals, 2*(1), 433.
    Orlikowski, W. J., & Scott, S. V. (2015). The Algorithm and the Crowd: Considering the Materiality of Service Innovation. *MIS Quarterly, 39*(1), 201–216.
20. Faulkner, P., & Runde, J. (2012). On Sociomateriality. In P. M. Leonardi, B. M. Nardi, & J. Kallinikos (Eds.), *Materiality and Organizing: Social Interaction in a Technological World.* Oxford: Oxford University Press, pp. 49–66.
21. McLuhan, M. (1964). *Understanding Media: The Extensions of Man.* London: Routledge.
    Strate, *ibid.*
22. Leonardi, P. M. (2010). Digital Materiality: How Artifacts without Matter, Matter. *First Monday, 15*(6).
23. Faulkner, P., & Runde, J. (2013). Technological Objects, Social Positions, and the Transformational Model of Social Reality. *MIS Quarterly, 37*(3), 803–818.
24. Mingers, J., Mutch, A., & Willcocks, L. P. (2013). Critical Realism in Information Systems Research. *MIS Quarterly, 37*(3), 795–802.
25. Espinosa, J. A., Slaughter, S. A., Kraut, R. E., & Herbsleb, J. D. (2007). Team Knowledge and Coordination in Geographically Distributed Software Development. *Journal of Management Information Systems, 24*(1), 135–169.
    Espinosa, J. A., Cummings, J. N., & Pickering, C. (2012). Time Separation, Coordination, and Performance in Technical Teams. *IEEE Transactions on Engineering Management, 59*(1), 91–103.
26. Postman, N. (1992). *Technopoly: The Surrender of Culture to Technology.* New York: Vintage, p. 14.
27. Dennis, A. R., George, J. F., Jessup, L. M., Nunamaker Jr, J. F., & Vogel, D. R. (1988). Information Technology to Support Electronic Meetings. *MIS Quarterly, 12*(4), 591–624.
28. Short et al, *ibid.*
29. Dennis, A. R., Wixom, B. H., & Vandenberg, R. J. (2001). Understanding Fit and Appropriation Effects in Group Support Systems via Meta-Analysis. *MIS Quarterly, 25*(2), 167–193.
    Karsten, H. (2003). Constructing Interdependencies with Collaborative Information Technology. *Computer Supported Cooperative Work: The Journal of Collaborative Computing, 12*, 437–464.
    McLeod, P. L., & Liker, J. K. (1992). Electronic Meeting Systems: Evidence from a Low Structure Environment. *Information Systems Research, 3*(3), 195–223.
    Ngwenyama, O. K., & Lee, A. S. (1997). Communication Richness in Electronic Mail: Critical Social Theory and the Contextuality of Meaning. *MIS Quarterly, 21*(2), 145–167.

Poole, M. S., Holmes, M., & DeSanctis, G. (1991). Conflict Management in a Computer-Supported Meeting Environment. *Management Science, 37*(8), 926–953.

Sambamurthy, V., & Poole, M. S. (1992). The Effects of Variations in Capabilities of GDSS Designs on Management of Cognitiive Conflict in Groups. *Information Systems Research, 3*(3), 224–251.

Tyran, C. K., Dennis, A. R., Vogel, D. R., & Nunamaker Jr., J. F. (1992). The Application of Electronic Meeting Technology to Support Strategic Management. *MIS Quarterly, 16*(3), 313–334.

Zigurs, I., & Buckland, B. K. (1998). A Theory of Task/Technology Fit and Group Support Systems Effectiveness. *MIS Quarterly, 22*(3), 313–334.

30. Dennis, A. R., George, J. F., Jessup, L. M., Nunamaker Jr, J. F., & Vogel, D. R. (1988). Information Technology to Support Electronic Meetings. *MIS Quarterly, 12*(4), 591–624.

31. Sørensen, C., & Kakihara, M. (2002). *Knowledge Discourses and Interaction Technology.* Paper presented at the 35th Hawaii International Conference on System Sciences (HICSS'02), Big Island, HI.

32. Barley, S. R. (1986). Technology as an Occasion for Structuring: Evidence from Observation of CT Scanners and the Social Order of Radiology Departments. *Administrative Science Quarterly, 31*(1), 78–108.

33. Yates, J., & Orlikowski, W. J. (1992). Genres of Organizational Communication: A Structurational Approach to Studying Communication and Media. *Academy of Management Review, 17*(2), 299–326.

34. Yates & Orlikowski, *ibid.*

Chapter 5

1. Borgmann, A. (1999). *Holding on to Reality: The Nature of Information at the Turn of the Century.* Chicago, IL: University of Chicago Press.

2. Shannon, C. E., & Weaver, W. 1963. *The Mathematical Theory of Communication.* Chicago, IL: University of Illinois Press.

3. Shannon & Weaver, *ibid.*, p. 22.

4. Shannon & Weaver, *ibid.*, p. 138.

5. Galbraith, J. R. (1974). Organization Design: An Information Processing View. *Interfaces, 4*(3), 28–36.
   Galbraith, J. R. (1977). *Organization Design.* Reading, MA: Addison-Wesley.

6. Daft, R. L., & MacIntosh, N. B. (1981). A Tentative Exploration into the Amount and Equivocality of Information Processing in Organizational Work Units. *Administrative Science Quarterly, 26*, 207–224.

7. Tushman, M. L., & Nadler, D. A. (1978). Information Processing as an Integrating Concept in Organizational Design. *Academy of Management Review, 3*(3), 613–624.

8. Tushman & Nadler, *ibid.*, p. 616.

9. Perrow, C. (1967). A Framework for the Comparative Analysis of Organizations. *American Sociological Review, 32*(2), 194–208.

10. "Z" is a mathematical specification notation that describes and models computer systems more clearly.

11. Ljungberg, F., & Sørensen, C. (2000). Overload: From Transaction to Interaction. In K. Braa, C. Sørensen, and B. Dahlbom (Eds.), *Planet Internet*. Lund: Studentlitteratur, pp. 113–136.

12. Buckley, W. (1968). Society as a Complex Adaptive System. In W. Buckley (Ed.), *Modern Systems Research for the Behavioral Scientist*. Chicago, IL: Aldine, pp. 490–513.

13. Tushman & Nadler, *ibid*.

14. Tushman & Nadler, *ibid*.

15. Grinter, R. E., Herbsleb, J. D., & Perry, D. E. (1999). *The Geography of Coordination: Dealing with Distance in R&D Work*. Paper presented at the GROUP'99, Phoenix, AZ.
Herbsleb, J. D., & Grinter, R. E. (1999). Architectures, Coordination, and Distance: Conway's Law and Beyond. *IEEE Software, 16*(5), 63–70.

16. Conway, M. E. (1968). How do Committees Invent? *Datamation, 14*(5), 28–31.

17. McCann, J. E., & Ferry, D. L. (1979). An Approach for Assessing and Managing Inter-unit Interdependence. *Academy of Management Review, 4*, 113–119.

18. Boehm, B., & Turner, R. (2004). *Balancing Agility and Discipline: A Guide for the Perplexed*. Boston, MA: Addison-Wesley.

19. Boh, W. F., Slaughter, S. A., & Espinosa, J. A. (2007). Learning from Experience in Software Development: A Multilevel Analysis. *Management Science, 53*(8), 1315–1331.

20. Espinosa, J. A., Kraut, R. E., Lerch, J. F., Slaughter, S. A., Herbsleb, J. D., & Audris, M. (2001). *Shared Mental Models and Coordination in Large-Scale, Distributed Software Development*. Paper presented at the 22nd International Conference on Information Systems (ICIS'01), New Orleans, LA.

21. Kotlarsky, J., & Oshri, I. (2005). Social Ties, Knowledge Sharing and Successful Collaboration in Globally Distributed System Development Projects. *European Journal of Information Systems, 14*, 37–48.

22. May, C., & Finch, T. (2009). Implementing, Embedding, and Integrating Practices: An Outline of Normalization Process Theory. *Sociology, 43*(5), 535–554.

23. Espinosa, J. A., Slaughter, S. A., Kraut, R. E., & Herbsleb, J. D. (2007). Team Knowledge and Coordination in Geographically Distributed Software Development. *Journal of Management Information Systems, 24*(1), 135–169.

24. Espinosa, J. A., & Carmel, E. (2003). The Impact of Time Separation on Coordination in Global Software Teams: A Conceptual Foundation. *Software Process Improvement and Practice, 8*, 249–266.
Espinosa, J. A., & Carmel, E. (2004). *The Effect of Time Separation on Coordination Costs in Global Software Teams: A Dyad Model*. Paper presented at the 37th Hawaiian International Conference on System Sciences, Big Island, HI.

25. Grinter et al., *ibid*.

26. Herbsleb & Grinter, *ibid*.

Chapter 6

1. Castells, M. (2009). *The Rise of the Network Society*. Oxford: Blackwell.
Castells, M. (2000). Materials for an Exploratory Theory of the Network Society. *British Journal of Sociology, 51*(1), 5–24.

2. Kallinikos, J. (2006). *The Consequences of Information: Institutional Implications of Technological Change.* Cheltenham: Edward Elgar, p. 9.

3. Grinter, R. E., Herbsleb, J. D., & Perry, D. E. (1999). *The Geography of Coordination: Dealing with Distance in R&D Work.* Paper presented at the GROUP'99, Phoenix, AZ.

4. Carmel, E., & Abbott, P. (2007). Why "Nearshore" Means That Distance Matters. *Communications of the ACM, 50*(10), 40–46.

5. Espinosa, J. A., Slaughter, S. A., Kraut, R. E., & Herbsleb, J. D. (2007). Team Knowledge and Coordination in Geographically Distributed Software Development. *Journal of Management Information Systems, 24*(1), 135–169.
   Sahay, S., Nicholson, B., & Krishna, S. (2003). *Global IT Outsourcing: Software Development Across Borders.* Cambridge: Cambridge University Press.

6. Cramton, C. D. (2001). The Mutual Knowledge Problem and Its Consequences for Dispersed Collaboration. *Organization Science, 12*(3), 346–371.

7. Niederman, F., & Tan, F. B. (2011). Managing Global IT Teams: Considering Cultural Dynamics. *Communications of the ACM, 54*(4), 24–27.

8. Damian, D., Lanubile, F., & Mallardo, T. (2008). On the Need for Mixed Media in Distributed Requirements Negotiations. *IEEE Transactions on Software Engineering, 34*(1), 116–132.

9. Oshri, I., van Fenema, P., & Kotlarsky, J. (2008). Knowledge Transfer in Globally Distributed Teams: The Role of Transactive Memory. *Information Systems Journal, 18*(6), 593–616.
   Avram, G. (2007). Knowledge Work Practices in Global Software Development. *The Electronic Journal of Knowledge Management, 5*(4), 347–356.

10. Oshri, I., Kotlarsky, J., & Willcocks, L. P. (2009). *The Handbook of Global Outsourcing and Offshoring.* Basingstoke: Palgrave Macmillan.

11. Zolin, R., Hinds, P. J., Fruchter, R., & Levitt, R. E. (2004). Interpersonal Trust in Cross-Functional Geographically Distributed Work. A Longitudinal Study. *Information and Organization, 14*, 1–26.

12. Mockus, A., & Weiss, D. M. (2001). Globalization by Chunking: A Quantitative Approach. *IEEE Software, 18*(2), 30–37.

13. Paasivaara, M., & Lassenius, C. (2003). Collaboration Practices in Global Inter-organizational Software Development Projects. *Software Process Improvement and Practice, 8*, 183–199.

14. Agnew, J. (2003). *Geopolitics: Re-Visioning World Politics* (2nd ed.). London: Routledge.

15. Rynasiewicz, R. (2014). Newton's Views on Space, Time and Motion. In E. N. Zalta (Ed.), *The Stanford Encyclopedia of Philosophy* (Vol. Summer). Stanford University. https://plato.stanford.edu/archives/sum2014/entries/newton-stm/. Accessed on April 17, 2019.

16. Isard, W. (1949). The General Theory of Location and Space-Economy. *Quarterly Journal of Economics, 63*(4), 476–506.
   Predöhl, A. (1928). The Theory of Location in Its Relation to General Economics. *Journal of Political Economy, 36*(3), 371–390.

17. O'leary, M., Orlikowski, W. J., & Yates, J. (2002). Distributed Work over the Centuries: Trust and Control in the Hudson's Bay Company, 1670–1826. In P. J. Hinds & S. Kiesler (Eds.), *Distributed Work.* Cambridge, MA: MIT Press, pp. 27–54.

18. O'Leary et al., *ibid.*, p. 29.

19. O'Leary et al., *ibid.*, p. 27 [a quotation from HBC's Executive Committee to one of its managers in North America, 1679 (in Rich 1948, p. 10)]
20. Agnew, *ibid.*
21. Agnew, *ibid.*
22. McLuhan, M. (1964/2001). *Understanding Media: The Extensions of Man*. London: Routledge.
23. Saunders, C., Rutkowski, A.-F., van Genuchten, M., Vogel, D. R., & Orrego, J. M. (2011). Virtual Space and Place: Theory and Test. *MIS Quarterly, 35*(4), 1079–1098.
24. Couclelis, H., & Gale, N. (1986). Space and Spaces. *Human Geography, 68*(1), 1–12.
25. Casey, E. S. (1993). *Getting Back into Place: Toward a Renewed Understanding of the Place-World*. Bloomington, IN: Indiana University Press.
    Casey, E. S. (1997). *The Fate of Place: A Philosophical History*. Berkeley, CA: University of California Press.
26. Relph, E. (1976). *Place and Placelessness*. London: Pion.
27. Cairncross, F. (1997). *The Death of Distance: How the Communications Revolution will Change Our Lives*. Boston, MA: Harvard Business School Press.
28. Bailey, D. E., Leonardi, P. M., & Barley, S. R. (2012). The Lure of the Virtual. *Organization Science, 23*(5), 1485–1504.
29. Oshri, I., van Fenema, P., & Kotlarsky, J. (2008). Knowledge Transfer in Globally Distributed Teams: The Role of Transactive Memory. *Information Systems Journal, 18*(6), 593–616.
30. Abler, R., Adams, J. S., & Gould, P. (1971). *Spatial Organization: The Geographer's View of the World*. Englewood Cliffs, NJ: Prentice Hall.
31. Agnew, *ibid.*, p. 317.
32. Agnew, *ibid.*, p. 317.
33. D'Mello, M., & Sahay, S. (2007). "I am Kind of a Nomad where I have to go Places and Places" ... Understanding Mobility, Place and Identity in Global Software Work from India. *Information and Organization, 17*(3), 162–192.
34. Feldman, M. P. (1994). *The Geography of Innovation*. Boston, MA: Kluwer.
35. Feldman, M. P. (2000). Location and Innovation: The New Economic Geography of Innovation, Spillovers and Agglomeration. In G. L. Clark, M. P. Feldman, & M. S. Gertler (Eds.), *The Oxford Handbook of Economic Geography*. Oxford: Oxford University Press, pp. 373–395.
36. Feldman, M. P. (2014). The Character of Innovative Places: Entrepreneurial Strategy, Economic Development, and Prosperity. *Small Business Economics, 43*(1), 9–20.
37. McLuhan, *ibid.*
38. Herbsleb, J. D., & Grinter, R. E. (1999). Architectures, Coordination, and Distance: Conway's Law and Beyond. *IEEE Software, 16*(5), 63–70.
39. Sack, R. D. (1992). *Place, Modernity, and the Consumer's World: A Relational Framework for Geographical Analysis*. Baltimore, MD: Johns Hopkins University Press.
40. Abowd, G. D., & Mynatt, E. D. (2000). Charting Past, Present, and Future Research in Ubiquitous Computing. *ACM Transactions on Computer-Human Interaction, 7*(1), 29–58.
41. Dourish, P., & Bell, G. (2011). *Divining a Digital Future: Mess and Mythology in Ubiquitous Computing*. Cambridge, MA: MIT Press.
42. Herbsleb & Grinter, *ibid.*
43. Balaji, S., & Brown, C. V. (2014). Lateral Coordination Mechanisms and the Moderating Role of Arrangement Characteristics in Information Systems Development Outsourcing. *Information Systems Research, 25*(4), 747–760.

44. Langer, N., Slaughter, S. A., & Mukhopadhyay, T. (2014). Project Managers' Practical Intelligence and Project Performance in Software Offshore Outsourcing: A Field Study. *Information Systems Research, 25*(2), 364–384.

45. Huang, P.-Y., Pan, S. L., & Ouyang, T. H. (2014). Developing Information Processing Capability for Operational Agility: Implications from a Chinese manufacturer. *European Journal of Information Systems, 23*(4), 462–480.

46. Levitt, R. E. (2012). The Virtual Design Team: Simulating how Organization Structure and Information Processing Tools affect Team Performance. *Journal of Organization Design, 1*(2), 14–41.

47. Griffith, T. L., & Neale, M. A. (2001). Information Processing in Traditional, Hybrid and Virtual Teams: From Nascent Knowledge to Transactive Memory. *Research in Organizational Behavior, 23,* 379–421.

48. Agnew, J. (2011). Space and Place. In J. Agnew & D. N. Livingstone (Eds.), *The SAGE Handbook of Geographical Knowledge.* London: Sage, pp. 316–330.

49. Parnas, D. L. (1972). On the Criteria to be Used in Decomposing Systems into Modules. *Communications of the ACM, 15*(2), 1053–1058.

50. Kotlarsky, J., & Oshri, I. (2005). Social Ties, Knowledge Sharing and Successful Collaboration in Globally Distributed System Development Projects. *European Journal of Information Systems, 14,* 37–48.

51. Mockus & Weiss, *ibid.*

52. Herbsleb, J. D., Mockus, A., Finholt, T. A., & Grinter, R. E. (2000). *Distance, Dependencies, and Delay in Global Collaboration.* Paper presented at the Conference on Computer Supported Cooperative Work, Philadelphia, PA.

53. Sarker, S., & Sarker, S. (2009). Exploring Agility in Distributed Information Systems Engineering Teams: An Interpretive Study in an Offshoring Context. *Information Systems Research, 20*(3), 440–461.

54. Feldman, M. P. (1994). *The Geography of Innovation.* Boston, MA: Kluwer.
Feldman, M. P. (2014). The Character of Innovative Places: Entrepreneurial Strategy, Economic Development, and Prosperity. *Small Business Economics, 43*(1), 9–20.

55. Schmidt, K., & Simone, C. (1996). Coordination Mechanisms: Towards a Conceptual Foundation of Computer Supported Cooperative Work Systems Design. *Computer Supported Cooperative Work: The Journal of Collaborative Computing, 5,* 155–200.

56. Harrison, S., & Dourish, P. (1996). *Re-Place-ing Space: The Roles of Place and Space in Collaborative Systems.* Paper presented at the ACM Conference on Computer-Supported Cooperative Work, Boston, MA.

57. Heidegger, M. (1962). *Being and Time.* Oxford: Blackwell.

58. Ó Conchúir, E., Ågerfalk, P. J., Olsson, H. H., & Fitzgerald, B. (2009). Global Software Development: Where Are the Benefits? *Communications of the ACM, 52*(8), 127–131.

59. Kotlarsky & Oshri, *ibid.*

60. Schmidt & Simone, *ibid.*

61. Schmidt & Simone, *ibid.*

62. Gibson, C. B., & Gibbs, J. L. (2006). Unpacking the Concept of Virtuality: The Effects of Geographic Dispersion, Electronic Dependence, Dynamic Structure, and National Diversity on Team Innovation. *Administrative Science Quarterly, 51*(3), 451–495.
Griffith, T. L., Sawyer, J. E., & Neale, M. A. (2003). Virtualness and Knowledge in teams: Managing the Love Triangle of Organizations, Individuals, and Information Technology. *MIS Quarterly, 27*(2), 265–287.

63. Levesque, L. L., Wilson, J. M., & Wholey, D. R. (2001). Cognitive Divergence and Shared Mental Models in Software Development Project Teams. *Journal of Organizational Behavior, 22*(2), 135–144.
64. Dourish, P. (2006). *Re-Space-ing Place: "Place" and "Space" Ten Years On*. Paper presented at the Computer Supported Cooperative Work (CSCW'06) Conference, Alberta, Canada.
65. Herbsleb et al., *ibid.*
66. Bailey et al., *ibid.*
67. Pfeffer, J., & Salancik, G. R. (1978). *The External Control of Organizations: A Resource Dependence Perspective*. Stanford, CA: Stanford University Press.
68. Levesque et al., *ibid.*
69. Leont'ev, A. N. (1981). *Problems of the Development of the Mind*. Moscow: Progress Publishers.
70. Lee, G., Delone, W., & Espinosa, J. A. (2006). Ambidextrous Coping Strategies in Globally Distributed Software Development Projects. *Communications of the ACM, 49*(10), 35–40.
71. Kotlarsky, J., Fenema, P. C. V., & Willcocks, L. P. (2008). Developing a Knowledge-Based Perspective on Coordination: The Case of Global Software Projects. *Information and Management, 45*, 96–108.
72. Espinosa et al., *ibid.*
73. Espinosa, J. A., & Carmel, E. (2003). The Impact of Time Separation on Coordination in Global Software Teams: A Conceptual Foundation. *Software Process Improvement and Practice, 8*, 249–266.
    Espinosa, J. A., & Carmel, E. (2004). *The Effect of Time Separation on Coordination Costs in Global Software Teams: A Dyad Model*. Paper presented at the 37th Hawaiian International Conference on System Sciences, Big Island, HI.
74. Grinter et al, *ibid.*
75. Herbsleb & Grinter, *ibid.*
    Ovaska, P. A., Rossi, M., & Marttiin, P. (2003). Architecture as A Coordination Tool in Multi-Site Software Development. *Software Process Improvement and Practice, 8*, 233–247.
76. Bailey et al., *ibid.*

Chapter 7

1. Cameron, K. S., & Quinn, R. E. (1988). Organizational Paradox and Transformation. In R. E. Quinn & K. S. Cameron (Eds.), *Paradox and Transformation: Toward a Theory of Change in Organization and Management*. Cambridge, MA: Ballinger, pp. 1–18.
2. Poole, M. S., & Van de Ven, A. H. (1989). Using Paradox to Build Management and Organization Theories. *Academy of Management Review, 14*(4), 562–578.
3. Sparr, J. L. (2018). Paradoxes in Organizational Change: The Crucial Role of Leaders' Sensegiving. *Journal of Change Management, 18*(2), 162–180.
4. Poole & Van de Ven, *ibid.*, p. 563.
5. Smith, W. K., & Lewis, M. W. (2011). Toward a Theory of Paradox: A Dynamic equilibrium Model of Organizing. *Academy of Management Review, 36*(2), 381–403.

6. Orton, J. D., & Weick, K. E. (1990). Loosely Coupled Systems: A Reconceptualization. *Academy of Management Review, 15*(2), 203–223.

7. Orton & Weick, *ibid.*

8. Mockus, A., & Weiss, D. M. (2001). Globalization by Chunking: A Quantitative Approach. *IEEE Software, 18*(2), 30–37.

9. Parnas, D. L. (1972). On the Criteria to be Used in Decomposing Systems into Modules. *Communications of the ACM, 15*(2), 1053–1058.

10. Orton & Weick, *ibid.*

11. Boynton, A. C., & Zmud, R. W. (1987). Information Technology Planning in the 1990s: Directions for Practice and Research. *MIS Quarterly, 11*(1), 59–71.
    Weick, K. E. (1982). Management of Organizational Change among Loosely Coupled Elements. In P. S. Goodman (Ed.), *Change in Organizations: New Perspectives on Theory and Research and Practice.* San Francisco, CA: Jossey-Bass, pp. 375–408.

12. Thompson, J. D. (1967). *Organizations in Action: Social Science Bases of Administrative Theory.* New Brunswick, NJ: Transactions Publishers, p. 11.

13. Benson, J. K. (1977). Organizations: A Dialectical View. *Administrative Science Quarterly, 22*(1), 1–21.

14. Herbsleb, J. D., & Grinter, R. E. (1999). Architectures, Coordination, and Distance: Conway's Law and Beyond. *IEEE Software, 16*(5), 63–70.

15. Conway, M. E. (1968). How do Committees Invent? *Datamation, 14*(5), 28–31.

16. Benson, *ibid.*

17. Rilla, N., & Squicciarini, M. (2011). R&D (Re)location and Offshore Outsourcing: A Management Perspective. *International Journal of Management Reviews, 13*(4), 393–413.

18. Scott, W. R. (2003). *Organizations: Rational, Natural and Open Systems* (5th ed.). Upper Saddle River, NJ: Prentice Hall.

19. Mockus & Weiss, *ibid.*
    Ó Conchúir, E., Ågerfalk, P. J., Olsson, H. H., & Fitzgerald, B. (2009). Global Software Development: Where Are the Benefits? *Communications of the ACM, 52*(8), 127–131.

20. Parnas, *ibid.*

21. Herbsleb, J. D., & Mockus, A. (2003). An Empirical Study of Speed and Communication in Globally Distributed Software Development. *IEEE Transactions on Software Engineering, 29*(6), 481–494.

22. Herbsleb & Grinter, *ibid.*

23. Espinosa, J., Slaughter, S., Kraut, R., & Herbsleb, J. (2007). Team Knowledge and Coordination in Geographically Distributed Software Development. *Journal of Management Information Systems, 24*(1), 135–169.
    Kotlarsky, J., & Oshri, I. (2005). Social Ties, Knowledge Sharing and Successful Collaboration in Globally Distributed System Development Projects. *European Journal of Information Systems, 14*(1), 37–48.
    Oshri, I., van Fenema, P., & Kotlarsky, J. (2008). Knowledge Transfer in Globally Distributed Teams: the Role of Transactive Memory. *Information Systems Journal, 18*(6), 593–616.

24. Ebert, C., & De Neve, P. (2001). Surviving Global Software Development. *IEEE Software, 18*(2), 62–69.

25. Sailer, K., & McCulloh, I. (2012). Social Networks and Spatial Configuration—How Office Layouts Drive Social Interaction. *Social Networks, 34*(1), 47–58.

26. Ó Conchúir et al., *ibid.*

27. Mockus & Weiss, *ibid.*

28. Sanchez, R., & Mahoney, J. T. (1996). Modularity, Flexibility and Knowledge Management in Product and Organization Design. *Strategic Management Journal, 17*(Winter Special Issue), 63–76.

29. Page-Jones, M. (1980). *The Practical Guide to Structures Systems Design*. New York: Yourdon Press.

30. Fitzgerald, B. (2000). Systems Development Methodologies: The Problem of Tenses. *Information Technology and People, 13*, 13–22.
    Kirsch, L. J. (1996). The Management of Complex Tasks in Organizations: Controlling the Systems Development Process. *Organization Science, 7*(1), 1–21.
    Malhotra, A., & Majchrzak, A. (2012). How Virtual Teams Use Their Virtual Workspace to Coordinate Knowledge. *ACM Transactions on Management Information Systems, 3*(1), 1–14.

31. Fitzgerald, *ibid.*

32. Kirsch, *ibid.*

33. Malhotra & Majchrzak, *ibid.*

34. Herbsleb & Grinter, *ibid.*

35. Maruping, L. M., Venkatesh, V., & Agarwal, R. (2009). A Control Theory Perspective on Agile Methodology Use and Changing User Requirements. *Information Systems Research, 20*(3), 377–399.

36. Sahay, S., Nicholson, B., & Krishna, S. (2003). *Global IT Outsourcing: Software Development Across Borders*. Cambridge: Cambridge University Press.

37. Ó Conchúir et al., *ibid.*
    Sahay et al., *ibid.*

38. Buckley, W. (1968). Society as a Complex Adaptive System. In W. Buckley (Ed.), *Modern Systems Research for the Behavioral Scientist*. Chicago, IL: Aldine, pp. 490–513.

39. Ó Conchúir et al., *ibid.*

40. Maruping et al., *ibid.*
    Nidumolu, S. R., & Subramani, M. R. (2004). The Matrix of Control: Combining Process and Structure Approaches to Managing Software Development. *Journal of Management Information Systems, 20*(3), 159–196.
    Gerwin, D., & Moffat, L. (1997). Withdrawal of Team Autonomy During Concurrent Engineering. *Management Science, 43*(9), 1275–1287.
    Sarker, S., & Sarker, S. (2009). Exploring Agility in Distributed Information Systems Development Teams: An Interpretive Study in an Offshoring Context. *Information Systems Research, 20*(3), 440–461.

41. Maruping et al., *ibid.*

42. Armstrong, D. J., & Cole, P. (2002). Managing Distances and Differences in Geographically Distributed Work Groups. In P. J. Hinds & S. Kiesler (Eds.), *Distributed Work*. London: MIT Press, pp. 167–186.

43. Maruping et al., *ibid.*

44. Alfaro, I. (2010). *National Diversity and Performance in Global Software Development Teams: The Role of Temporal Dispersion and Leadership*. Proceedings of the International Conference on Information Systems, St. Louis, MO. Paper 166.

45. Oshri, I., Kotlarsky, J., & Willcocks, L. P. (2007). Global Software Development: Exploring Socialization and Face-to-Face Meetings in Distributed Strategic Projects. *Journal of Strategic Information Systems, 16*(1), 25–49.

46. Zimmermann, A. (2010). Interpersonal Relationships in Transnational, Virtual Teams: Towards a Configurational Perspective. *International Journal of Managment Reviews*, *13*(1), 59–78.

47. Moe, N. B., & Šmite, D. (2008). Understanding a Lack of Trust in Global Software Teams: A Multiple-Case Study. *Software Process Improvement and Practice*, *13*(3), 217–231.

48. Kirsch, L. J. (1997). Portfolios of Control Modes and IS Project Management. *Information Systems Research*, *8*(3), 215–239.

49. Tushman, M. L., & Nadler, D. A. (1978). Information Processing as an Integrating Concept in Organizational Design. *Academy of Management Review*, *3*(3), 613–624.

50. Hinds, P. J., & McGrath, C. (2006). Structures that Work: Social Structure, Work Structure and Coordination Ease in Geographically Distributed Teams. Proceedings of the 20th Conference onComputer Supported Cooperative Work. Banff, Alberta, Canada, pp. 343–352.

51. Tushman, M. L. (1979). Work Characteristics and Subunit Communication Structure: A Contingency Analysis. *Administrative Science Quarterly*, *24*(1), 82–98.

52. Kotlarsky & Oshri, *ibid*.

53. Herbsleb, J. D. (2007). Global Software Engineering: The Future of Socio-technical Coordination. Proceedings of Future of Software Engineering 2007. Minneapolis, MN.

54. Sabherwal, R. (2003). The Evolution of Coordination in Outsourced Software Development Projects: A Comparison of Client and Vendor Perspectives. *Information and Organization*, *13*(3), 153–202.

55. Ovaska, P., Rossi, M., & Marttiin, P. (2003). Architecture as a Coordination Tool in Multi-site Software Development. *Software Process: Improvement and Practice*, *8*(4), 233–247.

56. Espinosa, J. A., & Carmel, E. (2004). The Impact of Time Separation on Coordination in Global Software Teams: A Conceptual Foundation. *Software Process: Improvement and Practice*, *8*, 249–266.

57. Oshri, I., van Fenema, P., & Kotlarsky, J. (2008). Knowledge Transfer in Globally Distributed Teams: the Role of Transactive Memory. *Information Systems Journal*, *18*(6), 593–616.

58. Kirsch, L. J. (1997). Portfolios of Control Modes and IS Project Management. *Information Systems Research*, *8*(3), 215–239.

59. Choudhury, V., & Sabherwal, R. (2003). Portfolios of Control in Outsourced Software Development Projects. *Information Systems Research*, *14*(3), 291–314.

60. Poole & Van de Ven, *ibid*.

61. Kanawattanachai, P., & Yoo, Y. (2007). The Impact of Knowledge Coordination on Virtual Team Performance over Time. *MIS Quarterly*, *31*(4), 783–808.

62. Kraut, R. E., Steinfield, C., Chan, A. P., Butler, B., & Hoag, A. (1999). Coordination and Virtualization: The Role of Electronic Networks and Personal Relationships. *Organization Science*, *10*(6), 722–740.

63. Lanubile, F., Mallardo, T., & Calefato, F. (2003). Tool support for geographically dispersed inspection teams. *Software Process: Improvement and Practice*, *8*(4), 217–231.

64. Nunamaker, J. F., Reinig, B. A., & Briggs, R. O. (2009). Principles for effective virtual teamwork. *Communications of the ACM*, *52*(4), 113.

65. Majchrzak, A., Rice, R. E., Malhotra, A., King, N., & Ba, S. (2000). Technology Adaptation: The Case of a Computer-Supported Inter-Organizational Virtual Team. *MIS Quarterly*, *24*(4), 569–600.

66. Bailey, D. E., Leonardi, P. M., & Barley, S. R. (2012). The Lure of the Virtual. *Organization Science*, *23*(5), 1485–1504.

Chapter 8

1. Gibson, C. B., & Gibbs, J. L. (2006). Unpacking the Concept of Virtuality: The Effects of Geographic dispersion, Electronic Dependence, Dynamic Structure, and National Diversity on Team Innovation. *Administrative Science Quarterly, 51*, 451–495.
   Schweitzer, L., & Duxbury, L. (2010). Conceptualizing and Measuring the Virtuality of Teams. *Information Systems Journal, 20*(3), 267–295.
2. Scott, W. R. (2003). *Organizations: Rational, Natural and Open Systems* (5th ed.). Upper Saddle River, NJ: Prentice Hall, p. 33.
3. Scott, *ibid.*, p. 83.
4. Hegel, G. W. F. (1807). *The Phenomenology of Spirit*. Oxford: Oxford University Press.
5. Lenin, V. I. (1965). On the Question of Dialectics. In V. I. Lenin (Ed.), *Collected Works* (Vol. 38). Moscow: Progress, pp. 220–222.
   Butler, C. (2012). *The Dialectical Method: A Treatise Hegel Never Wrote*. New York: Humanity Books.
6. Butler, *ibid.*, p. 69.
7. Marx, K., & Engels, F. (1887). *Capital Vol. 1*. Moscow: Progress.
8. Hirschheim, R., & Klein, H. K. (1989). Four Paradigms of Information Systems Development. *Communications of the ACM, 32*(10), 1199–1216.
   Hirschheim, R., & Klein, H. K. (1994). Realizing Emancipatory Principles in Information Systems Development: The Case for ETHICS. *MIS Quarterly, 18*(1), 83–109.
9. Bjerknes, G. (1991). Dialectical Reflection in Information Systems Development. *Scandinavian Journal of Information Systems, 3*, 55–77.
10. Carlo, J. L., Lyytinen, K., & Boland Jr., R. J. (2012). Dialectics of Collective Minding: Contradictory Appropriations of Information Technology in a High-Risk Project. *MIS Quarterly, 36*(4), 1081–1108.
11. Robey, D., Ross, J. W., & Boudreau, M.-C. (2002). Learning to Implement Enterprise Systems: An Exploratory Study of the Dialectics of Change. *Journal of Management Information Systems, 19*(1), 17–46.
12. Myers, M. D. (1994). Dialectical Hermeneutics: A Theoretical Framework for the Implementation of Information Systems. *Information Systems Journal, 5*, 51–70.
13. Swanson, E. B. (1988). *Information System Implementation: Bridging the Gap between Design and Utilization*. Homewood, IL: Irwin.
14. Keen, P. G. W. (1981). Information Systems and Organizational Change. *Communications of the ACM, 24*(1), 24–33.
15. Thompson, J. D. (1967). *Organizations in Action* (2nd ed.). New Brunswick, NJ: Transactions Publishers.
16. March, J. G., & Simon, H. (1958). *Organizations*. New York: Wiley.
    Thompson, *ibid.*
17. Nidumolu, S. R., & Subramani, M. R. (2004). The Matrix of Control: Combining Process and Structure Approaches to Managing Software Development. *Journal of Management Information Systems, 20*(3), 159–196.
18. Herbsleb, J. D., & Grinter, R. E. (1999). Architectures, Coordination, and Distance: Conway's Law and Beyond. *IEEE Software, 16*(5), 63–70.
19. Ovaska, P., Rossi, M., & Marttiin, P. (2003). Architecture as a Coordination Tool in Multi-site Software Development. *Software Process Improvement and Practice, 8*, 233–247.

20. Kotlarsky, J., Oshri, I., & Hillegersberg, J. (2007). Globally Distributed Component-Based Software Development: An Exploratory Study of Knowledge Management and Work Division. *Journal of Information Technology, 22*(2), 161–173.

21. Kirsch, L. J. (2004). Deploying Common Systems Globally: The Dynamics of Control. *Information Systems Research, 15*(4), 374–395.
    Kirsch, L. J., Sambamurthy, V., Ko, D., & Purvis, R. L. (2002). Controlling Information Systems Development Projects: The View from the Client. *Management Science, 48*(4), 484–498.

22. Maruping, L. M., Venkatesh, V., & Agarwal, R. (2009). A Control Theory Perspective on Agile Methodology Use and Changing User Requirements. *Information Systems Research, 20*(3), 377–399.

23. Lee, G., Espinosa, J. A., & DeLone, W. H. (2013). Task Environment Complexity, Global Team Dispersion, Process Capabilities, and Coordination in Software Development. *IEEE Transactions on Software Engineering, 39*(12), 1753–1769.

24. Herbsleb, J. D., & Moitra, D. (2001). Global Software Development. *IEEE Software, 18*(2), 17–20.

25. Brooks, F. P. (1995). *The Mythical Man-Month: Essays on Software Engineering.* New York: Addison-Wesley.

26. Lee et al., *ibid.*

27. Thompson, *ibid.*

28. Parnas, D. L. (1972). On the Criteria to be Used in Decomposing Systems into Modules. *Communications of the ACM, 15*(2), 1053–1058.

29. Mockus, A., & Weiss, D. M. (2001). Globalization by Chunking: A Quantitative Approach. *IEEE Software, 18*(2), 30–37.

30. Maruping et al., *ibid.*
    Nidumolu & Subramani, *ibid.*
    Gerwin, D., & Moffat, L. (1997). Withdrawal of Team Autonomy During Concurrent Engineering. *Management Science, 43*(9), 1275–1287.
    Sarker, S., & Sarker, S. (2009). Exploring Agility in Distributed Information Systems Development Teams: An Interpretive Study in an Offshoring Context. *Information Systems Research, 20*(3), 440–461.

31. Maruping et al., *ibid.*

32. Orlikowski, W. J. (2002). Knowing in Practice: Enacting a Collective Capability in Distributed Organizing. *Organization Science, 13*(3), 249–273.

33. Cunha, M. P. E., & Cunha, J. V. D. (2001). Managing Improvisation in Cross Cultural Teams. *International Journal of Cross Cultural Management, 1*(2), 187–208.

34. Lee, G., Delone, W., & Espinosa, J. A. (2006). Ambidexterous Coping Strategies in Globally Distributed Software Development Projects. *Communications of the ACM, 49*(10), 35–40.

35. Lee, G., Espinosa, J. A., & DeLone, W. H. (2013). Task Environment Complexity, Global Team Dispersion, Process Capabilities, and Coordination in Software Development. *IEEE Transactions on Software Engineering, 39*(12), 1753–1769.

36. Espinosa, J. A., Slaughter, S. A., Kraut, R. E., & Herbsleb, J. D. (2007). Team Knowledge and Coordination in Geographically Distributed Software Development. *Journal of Management Information Systems, 24*(1), 135–169.

37. Mani, D., Srikanth, K., & Bharadwaj, A. (2014). Efficacy of R&D Work in Offshore Captive Centers: An Empirical Study of Task Characteristics, Coordination Mechanisms, and Performance. *Information Systems Research, 25*(4), 846–864.

38. Alfaro, I. (2010). *National Diversity and Performance in Global Software Development Teams: The Role of Temporal Dispersion and Leadership.* Paper presented at the International Conference on Information Systems, St. Louis, MO. Paper 166.
   Alfaro, I. (2014). Helping Global Software-Development Teams to Overcome the Challenges of Temporal Dispersion and National Diversity: The Importance of Leadership Roles. In F. Rowe & D. Te'eni (Eds.), *Innovation and IT in an International Context.* London: Palgrave Macmillan, pp. 189–210.

39. Oshri, I., Kotlarsky, J., & Willcocks, L. P. (2007). Global Software Development: Exploring Socialization and Face-to-Face Meetings in Distributed Strategic Projects. *Journal of Strategic Information Systems, 16*(1), 25–49.

40. Moe, N. B., & Smite, D. (2008). Understanding a Lack of Trust in Global Software Teams: A Multiple-Case Study. *Software Process Improvement and Practice, 13*(3), 217–231.

41. Kirsch, L. J. (1997). Portfolios of Control Modes and IS Project Management. *Information Systems Research, 8*(3), 215–239.

42. Tushman, M. L. (1979). Work Characteristics and Subunit Communication Structure: A Contingency Analysis. *Administrative Science Quarterly, 24*(1), 82–98.

43. Mockus & Weiss, *ibid.*

44. Carmel, E. (1999). *Global Software Teams: Collaborating Across Borders and Time Zones.* Upper Saddle River, NJ: Prentice Hall.

45. Aversano, L., Lucia, A. D., Gaeta, M., Ritrovato, P., Stefanucci, S., & Villani, M. L. (2004). Managing Coordination and Cooperation in Distributed Software Processes: The GENESIS Environment. *Software Process Improvement and Practice, 9*(4), 239–263.

46. Kotlarsky, J., & Oshri, I. (2005). Social Ties, Knowledge Sharing and Successful Collaboration in Globally Distributed System Development Projects. *European Journal of Information Systems, 14*, 37–48.

47. Couclelis, H., & Gale, N. (1986). Space and Spaces. *Human Geography, 68*(1), 1–12.
   Saunders, C., Rutkowski, A.-F., van Genuchten, M., Vogel, D. R., & Orrego, J. M. (2011). Virtual Space and Place: Theory and Test. *MIS Quarterly, 35*(4), 1079–1098.

48. Curtis, B., Krasner, H., & Iscoe, N. (1988). A Field Study of the Software Design Process for Large Systems. *Communications of the ACM, 31*(11), 1268–1287.

49. Borgmann, A. (1999). *Holding on to Reality: The Nature of Information at the Turn of the Century.* Chicago, IL: University of Chicago Press.

50. Evans, P., & Wurster, T. S. (2000). *Blown to Bits: How the New Economics of Information Transforms Strategy.* Boston, MA: HBR Press.

51. Kallinikos, J. (2012). Form, Function, and Matter: Crossing the Border of Materiality. In P. M. Leonardi, B. M. Nardi & J. Kallinikos (Eds.), *Materiality and Organizing: Social Interaction in a Technological World.* Oxford: Oxford University Press, p. 77.

52. Kallinikos, *ibid.*, p. 81.

53. Dubé, L., & Robey, D. (2008). Surviving the Paradoxes of Virtual Teamwork. *Information Systems Journal, 19*(1), 3–30.

54. Sarker, S., & Sahay, S. (2004). Implications of Space and Time for Distributed Work: An Interpretive Study of US-Norwegian Systems Development Teams. *European Journal of Information Systems, 13*, 3–20.

55. Jehn, K. A. (1997). A Qualitative Analysis of Conflict Types and Dimensions in Organizational Groups. >*Administrative Science Quarterly, 42* (3), 530–557.

56. Holmström, H., Fitzgerald, B., Agerfalk, P., & O Conchuir, E. (2008). Agile Practices Reduce Distance in Global Software Development. *Information Systems Management, 23*(3), 7–18.

Smite, D., Moe, N. B., & Ågerfalk, P. J. (Eds.). (2010). *Agility Across Time and Space: Implementing Agile Methods in Global Software Projects.* Berlin: Springer.

Schwaber, K. (2004). *Agile Project Management with Scrum.* Redmond, WA: Microsoft Press.

Jalali, S., & Wohlin, C. (2012). Global Software Engineering and Agile Practices: A Systematic Review. *Journal of Software Maintenance and Evolution: Research and Practice, 24*(6), 643–659.

57. Carmel, E., Espinosa, J. A., & Dubinsky, Y. (2010). "Follow the Sun" Workflow in Global Software Development. *Journal of Management Information Systems, 27*(1), 17–38.

    Colazo, J. A., & Fang, Y. (2010). Following the Sun: Temporal Dispersion and Performance in Open Source Software Project Teams. *Journal of the Association for Information Systems, 11*(11/12), 684–707.

58. Krishna, S., Sahay, S., & Walsham, G. (2004). Managing Cross-cultural Issues in Global Software Outsourcing. *Communications of the ACM, 47*(4), 62–66.

    Nicholson, B., & Sahay, S. (2001). Some Political and Cultural Issues in the Globalisation of Software Development: Case Experience from Britain and India. *Information and Organization, 11*, 25–43.

    Niederman, F., & Tan, F. B. (2011). Managing Global IT Teams: Considering Cultural Dynamics. *Communications of the ACM, 54*(4), 24–27.

59. Sarker & Sahay, *ibid.*

Chapter 9

1. Dubé, L., & Paré, G. (2003). Rigor in Information Systems Positivist Case Research: Current Practices, Trends, and Recommendations. *MIS Quarterly, 27*(4), 597–636.

   Leonardi, P. M. (2011). When Flexible Routines Meet Flexible Technologies: Affordance, Constraint, and the Imbrication of Human and Material Agencies. *MIS Quarterly, 35*(1), 147–167.

2. Avgerou, C. (2013). Social Mechanisms for Causal Explanation in Social Theory Based IS Research. *Journal of the Association for Information Systems, 14*(8), 399–419.

3. Tsoukas, H. (1989). The Validity of Idiographic Research Explanations. *Academy of Management Review, 14*(4), 551–561.

4. Lindberg, A., Berente, N., Gaskin, J., & Lyytinen, K. (2016). Coordinating Interdependencies in Online Communities: A Study of an Open Source Software Project. *Information Systems Research, 15*(2), 751–772.

   Srikanth, K., & Puranam, P. (2014). The Firm as a Coordination System: Evidence from Software Services Offshoring. *Organization Science, 25*(4), 1253–1271.

   Majchrzak, A., Rice, R. E., Malhotra, A., King, N., & Ba, S. (2000). Technology Adaptation: The Case of a Computer-Supported Inter-Organizational Virtual Team. *MIS Quarterly, 24*(4), 569–600.

   Oshri, I., van Fenema, P., & Kotlarsky, J. (2008). Knowledge Transfer in Globally Distributed Teams: The Role of Transactive Memory. *Information Systems Journal, 18*(6), 593–616.

   Espinosa, J. A., Cummings, J. N., & Pickering, C. (2012). Time Separation, Coordination, and Performance in Technical Teams. *IEEE Transactions on Engineering Management, 59*(1), 91–103.

5. Europe, Middle East, and Africa.

Chapter 10

1. Leonardi, P. M., Nardi, B. M., & Kallinikos, J. (Eds.). (2012). *Materiality and Organizing: Social Interaction in a Technological World.* Oxford: Oxford University Press.
2. Carmel, E. (1999). *Global Software Teams: Collaborating Across Borders and Time Zones.* Upper Saddle River, NJ: Prentice Hall.
   Kotlarsky, J., & Oshri, I. (2005). Social Ties, Knowledge Sharing and Successful Collaboration in Globally Distributed System Development Projects. *European Journal of Information Systems, 14,* 37–48.
   Majchrzak, A., Rice, R. E., Malhotra, A., King, N., & Ba, S. (2000). Technology Adaptation: The Case of a Computer-Supported Inter-Organizational Virtual Team. *MIS Quarterly, 24*(4), 569–600.
   Mockus, A., & Weiss, D. M. (2001). Globalization by Chunking: A Quantitative Approach. *IEEE Software, 18*(2), 30–37.
3. Majchrzak et al., *ibid.*
4. Malhotra, A., & Majchrzak, A. (2012). How Virtual Teams Use Their Virtual Workspace to Coordinate Knowledge. *ACM Transactions on Management Information Systems, 3*(1), 1–14.
5. Majchrzak et al., *ibid.*
6. Malone, T. W., & Crowston, K. (1994). The Interdisciplinary Study of Coordination. *ACM Computing Surveys, 26*(1), 87–119.
7. Herbsleb, J. D., & Grinter, R. E. (1999). *Splitting the Organization and Integrating the Code: Conway's Law Revisited.* Paper presented at the ICSE'99, Los Angeles, CA, pp. 85–95.
8. Herbsleb, J. D., Mockus, A., Finholt, T. A., & Grinter, R. E. (2000). *Distance, Dependencies, and Delay in Global Collaboration.* Proceedings of the Conference on Computer Supported Cooperative Work, Philadelphia, PA.
9. Material cause: that out of which a thing comes to be, and which persists.
   Formal cause: the statement of essence, the account of what-it-is- to-be, and the parts of the account.
   Efficient cause: the primary source of the change.
   Final cause: the end (telos), that for the sake of which a thing is done.
10. Falcon, A. (2015). Aristotle on Causality. In *The Stanford Encyclopedia of Philosophy.* Retrieved on January 22, 2019, from https://plato.stanford.edu/archives/spr2015/entries/aristotle-causality.
11. Thompson, J. D. (1956). On Building an Administrative Science. *Administrative Science Quarterly, 1*(1), 102–111.
12. Poole, M. S., & Van de Ven, A. H. (1989). Using Paradox to Build Management and Organization Theories. *Academy of Management Review, 14*(4), 562–578.
13. Lee, G., Delone, W., & Espinosa, J. A. (2006). Ambidexterous Coping Strategies in Globally Distributed Software Development Projects. *Communications of the ACM, 49*(10), 35–40.
14. Herbsleb, J. D., & Grinter, R. E. (1999). Architectures, Coordination, and Distance: Conway's Law and Beyond. *IEEE Software, 16*(5), 63–70.
15. Orlikowski, W. J. (2002). Knowing in Practice: Enacting a Collective Capability in Distributed Organizing. *Organization Science, 13*(3), 249–273.

16. Cunha, M. P. E., & Cunha, J. V. D. (2001). Manging Improvisation in Cross Cultural Teams. *International Journal of Cross Cultural Management, 1*(2), 187–208.

17. Gilson, L. L., Maynard, M. T., Young, N. C. J., Vartianinen, M., & Hakonen, M. (2015). Virtual Teams Research: 10 Years, 10 Themes, and 10 Opportunities. *Journal of Management, 41*(5), 1313–1337.

18. Leonardi et al., *ibid.*

19. March, J. G. (2004). Parochialism in the Evolution of a Research Community: The Case of Organization Studies. *Management and Organization Review, 1*(1), 5–22.

20. Gregor, S. (2006). The Nature of Theory in Information Systems. *MIS Quarterly, 30*(3), 611–642.

Appendix

Aversano, L., Lucia, A. D., Gaeta, M., Ritrovato, P., Stefanucci, S., & Villani, M. L. (2004). Managing Coordination and Cooperation in Distributed Software Processes: The GENESIS Environment. *Software Process Improvement and Practice, 9*(4), 239–263.

Bjørn, P., Søderberg, A. M., & Krishna, S. (2019). Translocality in Global Software Development: the Dark Side of Global Agile. *Human-Computer Interaction, 34*(2), 174–203.

Colazo, J. A., & Fang, Y. (2010). Following the Sun: Temporal Dispersion and Performance in Open Source Software Project Teams. *Journal of the Association for Information Systems, 11*(11/12), 684–707.

D'Aubeterre, F., Singh, R., & Iyer, L. (2008). Secure Activity Resource Coordination: Empirical Evidence of Enhanced Security Awareness in Designing Secure Business Processes. *European Journal of Information Systems, 17*(5), 528–542.

David, G. C., Chand, D., Newell, S., & Resende-Santos, J. (2008). Integrated Collaboration Across Distributed Sites: The Perils of Process and the Promise of Practice. *Journal of Information Technology, 23*(1), 44–54.

Dennis, A. R., Wixom, B. H., & Vandenberg, R. J. (2001). Understanding Fit and Appropriation Effects in Group Support Systems via Meta-Analysis. *MIS Quarterly, 25*(2), 167–193.

Du, R., Ai, S., Abbott, P., & Zheng, Y. (2011). Contextual Factors, Knowledge Processes and Performance in Global Sourcing of IT Services: An Investigation in China. *Journal of Global Information Management, 19*(2), 1–26.

Esbensen, M., & Bjørn, P. (2014). Routine and Standardization in Global Software Development. *Proceedings of the 18th International Conference on Supporting Group Work*, Sanibel Island, FL, 12–23.

Espinosa, J. A., & Carmel, E. (2003). The Impact of Time Separation on Coordination in Global Software Teams: A Conceptual Foundation. *Software Process Improvement and Practice, 8*, 249–266.

Espinosa, J. A., & Carmel, E. (2004). *The Effect of Time Separation on Coordination Costs in Global Software Teams: A Dyad Model*. Paper presented at the 37th Hawaiian International Conference on System Sciences, Big Island, HI.

Espinosa, J. A., Cummings, J. N., & Pickering, C. (2012). Time Separation, Coordination, and Performance in Technical Teams. *IEEE Transactions on Engineering Management, 59*(1), 91–103.

Espinosa, J. A., Slaughter, S. A., Kraut, R. E., & Herbsleb, J. D. (2007a). Familiarity, Complexity, and Team Performance in Geographically Distributed Software Development. *Organization Science, 18*(4), 613–630.

Espinosa, J. A., Slaughter, S. A., Kraut, R. E., & Herbsleb, J. D. (2007b). Team Knowledge and Coordination in Geographically Distributed Software Development. *Journal of Management Information Systems, 24*(1), 135–169.

Ghobadi, S. (2015). What Drives Knowledge Sharing in Software Development Teams: A Literature Review and Classification Framework. *Information and Management, 52*(1), 82–97.

Grinter, R. E., Herbsleb, J. D., & Perry, D. E. (1999). *The Geography of Coordination: Dealing with Distance in R&D Work.* Paper presented at GROUP'99, Phoenix, AZ.

Herbsleb, J. D., & Grinter, R. E. (1999). Architectures, Coordination, and Distance: Conway's Law and Beyond. *IEEE Software, 16*(5), 63–70.

Herbsleb, J. D., & Mockus, A. (2003). An Empirical Study of Speed and Communication in Globally Distributed Software Development. *IEEE Transactions on Software Engineering, 29*(6), 481–494.

Herbsleb, J. D., Mockus, A., Finholt, T. A., & Grinter, R. E. (2000). *Distance, Dependencies, and Delay in Global Collaboration.* Paper presented at the Conference on Computer Supported Cooperative Work, Philadelphia, PA.

Huang, W., Wei, K.-K., & Lim, J. (2003). Using a GSS to Support Virtual Teambuilding: A Theoretical Framework. *Journal of Global Information Management.*

Im, H.-G., Yates, J., & Orlikowski, W. (2005). Temporal Coordination through Communication: Using Genres in a Virtual Start-up Organization. *Information Technology and People, 18*(2), 89–119.

Karsten, H. (2003). Constructing Interdependencies with Collaborative Information Technology. *Computer Supported Cooperative Work: The Journal of Collaborative Computing, 12*, 437–464.

Koppman, S., & Gupta, A. (2014). Navigating the Mutual Knowledge Problem: A Comparative Case Study of Distributed Work. *Information Technology and People, 27*(1), 83–105.

Kotlarsky, J., Fenema, P. C. V., & Willcocks, L. P. (2008). Developing a Knowledge-Based Perspective on Coordination: The Case of Global Software Projects. *Information and Management, 45*, 96–108.

Kotlarsky, J., & Oshri, I. (2005). Social Ties, Knowledge Sharing and Successful Collaboration in Globally Distributed System Development Projects. *European Journal of Information Systems, 14*, 37–48.

Kotlarsky, J., Oshri, I., & Hillegersberg, J. (2007). Globally Distributed Component-Based Software Development: An Exploratory Study of Knowledge Management and Work Division. *Journal of Information Technology, 22*(2), 161–173.

Lee, G., Espinosa, J. A., & DeLone, W. H. (2013). Task Environment Complexity, Global Team Dispersion, Process Capabilities, and Coordination in Software Development. *IEEE Transactions on Software Engineering, 39*(12), 1753–1769.

Lindberg, A., Berente, N., Gaskin, J., & Lyytinen, K. (2016). Coordinating Interdependencies in Online Communities: A Study of an Open Source Software Project. *Information Systems Research, 15*(2), 751–772.

Mani, D., Srikanth, K., & Bharadwaj, A. (2014). Efficacy of R&D Work in Offshore Captive Centers: An Empirical Study of Task Characteristics, Coordination Mechanisms, and Performance. *Information Systems Research, 25*(4), 846–864.

Mattarelli, E., & Gupta, A. (2009). Offshore-Onsite Subgroup Dynamics in Globally Distributed Teams. *Information Technology and People, 22*(3), 242–269.

McLeod, P. L., & Liker, J. K. (1992). Electronic Meeting Systems: Evidence from a Low Structure Environment. *Information Systems Research*, *3*(3), 195–223.

Nguyen-Duc, A., Cruzes, D. S., & Conradi, R. (2015). The Impact of Global Dispersion on Coordination, Team Performance and Software Quality—A Systematic Literature Review. *Information and Software Technology*, *57*(1), 277–294.

Ngwenyama, O. K. (1998). Groupware, Social Action and Organizational Emergence: On the Process Dynamics of Computer Mediated Distributed Work. *Accounting, Management and Information Technology*, *8*, 127–146.

Niederman, F., & Tan, F. B. (2011). Managing Global IT Teams: Considering Cultural Dynamics. *Communications of the ACM*, *54*(4), 24–27.

Oshri, I., van Fenema, P., & Kotlarsky, J. (2008). Knowledge Transfer in Globally Distributed Teams: The Role of Transactive Memory. *Information Systems Journal*, *18*(6), 593–616.

Ovaska, P. A., Rossi, M., & Marttiin, P. (2003). Architecture as a Coordination Tool in Multi-site Software Development. *Software Process Improvement and Practice*, *8*, 233–247.

Paasivaara, M., & Lassenius, C. (2003). Collaboration Practices in Global Inter-Organizational Software Development Projects. *Software Process Improvement and Practice*, *8*(4), 183–199.

Poole, M. S., Holmes, M., & DeSanctis, G. (1991). Conflict Management in a Computer-Supported Meeting Environment. *Management Science*, *37*(8), 926–953.

Sabherwal, R. (2003). The Evolution of Coordination in Outsourced Software Development Projects: A Comparison of Client and Vendor Perspectives. *Information and Organization*, *13*, 153–202.

Sambamurthy, V., & Poole, M. S. (1992). The Effects of Variations in Capabilities of GDSS Designs on Management of Cognitiive Conflict in Groups. *Information Systems Research*, *3*(3), 224–251.

Srikanth, K., & Puranam, P. (2011). Integrating Distributed Work: Comparing Task Design, Communication, and Tacit Coordination Mechanisms. *Strategic Management Journal*, *32*(8), 849–875.

Srikanth, K., & Puranam, P. (2014). The Firm as a Coordination System: Evidence from Software Services Offshoring. *Organization Science*, *25*(4), 1253–1271.

Thomas, D., & Bostrom, R. (2010). Team Leader Strategies for Enabling Collaboration Technology Adaptation: Team Technology Knowledge to Improve Globally Distributed Systems Development Work. *European Journal of Information Systems*, *19*(2), 223–237.

Tyran, C. K., Dennis, A. R., Vogel, D. R., & Nunamaker Jr., J. F. (1992). The Application of Electronic Meeting Technology to Support Strategic Management. *MIS Quarterly*, *16*(3), 313–334.

Zahedi, M., Shahin, M., & Babar, M. A. (2016). A Systematic Review of Knowledge Sharing Challenges and Practices in Global Software Development. *International Journal of Information Management*, *36*(6).

Zigurs, I., & Buckland, B. K. (1998). A Theory of Task/Technology Fit and Group Support Systems Effectiveness. *MIS Quarterly*, *22*(3), 313–334.

Zimmermann, A., & Ravishankar, M. N. (2014). Knowledge Transfer in IT Offshoring Relationships: The Roles of Social Capital, Efficacy and Outcome Expectations. *Information Systems Journal*, *24*(2), 167–202.

Index